# Extending R

# Chapman & Hall/CRC
# The R Series

## Series Editors

**John M. Chambers**
Department of Statistics
Stanford University
Stanford, California, USA

**Torsten Hothorn**
Division of Biostatistics
University of Zurich
Switzerland

**Duncan Temple Lang**
Department of Statistics
University of California, Davis
Davis, California, USA

**Hadley Wickham**
RStudio
Boston, Massachusetts, USA

## Aims and Scope

This book series reflects the recent rapid growth in the development and application of R, the programming language and software environment for statistical computing and graphics. R is now widely used in academic research, education, and industry. It is constantly growing, with new versions of the core software released regularly and more than 7,000 packages available. It is difficult for the documentation to keep pace with the expansion of the software, and this vital book series provides a forum for the publication of books covering many aspects of the development and application of R.

The scope of the series is wide, covering three main threads:

- Applications of R to specific disciplines such as biology, epidemiology, genetics, engineering, finance, and the social sciences.
- Using R for the study of topics of statistical methodology, such as linear and mixed modeling, time series, Bayesian methods, and missing data.
- The development of R, including programming, building packages, and graphics.

The books will appeal to programmers and developers of R software, as well as applied statisticians and data analysts in many fields. The books will feature detailed worked examples and R code fully integrated into the text, ensuring their usefulness to researchers, practitioners and students.

# Published Titles

**Extending R**, *John M. Chambers*

**Stated Preference Methods Using R**, *Hideo Aizaki, Tomoaki Nakatani, and Kazuo Sato*

**Using R for Numerical Analysis in Science and Engineering**, *Victor A. Bloomfield*

**Event History Analysis with R**, *Göran Broström*

**Computational Actuarial Science with R**, *Arthur Charpentier*

**Statistical Computing in C++ and R**, *Randall L. Eubank and Ana Kupresanin*

**Basics of Matrix Algebra for Statistics with R**, *Nick Fieller*

**Reproducible Research with R and RStudio, Second Edition**, *Christopher Gandrud*

**R and MATLAB®***David E. Hiebeler*

**Statistics in Toxicology Using R** *Ludwig A. Hothorn*

**Nonparametric Statistical Methods Using R**, *John Kloke and Joseph McKean*

**Displaying Time Series, Spatial, and Space-Time Data with R**, *Oscar Perpiñán Lamigueiro*

**Programming Graphical User Interfaces with R**, *Michael F. Lawrence and John Verzani*

**Analyzing Sensory Data with R**, *Sébastien Lê and Theirry Worch*

**Parallel Computing for Data Science: With Examples in R, C++ and CUDA**, *Norman Matloff*

**Analyzing Baseball Data with R**, *Max Marchi and Jim Albert*

**Growth Curve Analysis and Visualization Using R**, *Daniel Mirman*

**R Graphics, Second Edition**, *Paul Murrell*

**Introductory Fisheries Analyses with R**, *Derek H. Ogle*

**Data Science in R: A Case Studies Approach to Computational Reasoning and Problem Solving**, *Deborah Nolan and Duncan Temple Lang*

**Multiple Factor Analysis by Example Using R**, *Jérôme Pagès*

**Customer and Business Analytics: Applied Data Mining for Business Decision Making Using R**, *Daniel S. Putler and Robert E. Krider*

**Implementing Reproducible Research**, *Victoria Stodden, Friedrich Leisch, and Roger D. Peng*

**Graphical Data Analysis with R**, *Antony Unwin*

**Using R for Introductory Statistics, Second Edition**, *John Verzani*

**Advanced R**, *Hadley Wickham*

**Dynamic Documents with R and knitr, Second Edition**, *Yihui Xie*

# Extending R

**John M. Chambers**

Stanford University, California, USA

CRC Press
Taylor & Francis Group
Boca Raton   London   New York

CRC Press is an imprint of the
Taylor & Francis Group, an **informa** business

A CHAPMAN & HALL BOOK

CRC Press
Taylor & Francis Group
6000 Broken Sound Parkway NW, Suite 300
Boca Raton, FL 33487-2742

© 2016 by Taylor & Francis Group, LLC
CRC Press is an imprint of Taylor & Francis Group, an Informa business

No claim to original U.S. Government works

Printed on acid-free paper
Version Date: 20160421

International Standard Book Number-13: 978-1-4987-7571-7 (Paperback)

**Visit the Taylor & Francis Web site at**
**http://www.taylorandfrancis.com**

**and the CRC Press Web site at**
**http://www.crcpress.com**

*In memory*
*of my dear soulmate*

# Contents

# Preface

The role of R is not just as a programming language or an interactive environment for data analysis. It provides both of these, but I believe its most transformative role has been when individuals and groups add their own contribution, *extending* R. These extensions include many thousands of packages and other useful aids to learning about data and sharing the results.

This book aims to support current and future contributors by providing ideas and tools.

Extensions to R can be modest or ambitious. They may involve nothing more than adding some software to handle the tasks typically encountered; integrating R with your work flow; making your research results available for others to apply. At the other extreme, R may be taken on as part of an ambitious project with a challenging set of goals. Whatever the scale, you have something of value in the software you are creating. Organizing it effectively and using appropriate techniques will improve the programming process and the end result.

A few concepts, what I'm calling the three fundamental principles, are guides to designing and organizing extensions to R: objects, functions and interfaces. That users *will* extend the system has always been a major part of the design, going back to the original S language on which R is based. That came from a statistics research community, where new ideas were our main concern. The object-based, functional form of R makes new computations and new types of data natural. Interfaces to other software are an integral part of programming with R, and always have been. Understanding objects, functions and interfaces in extending R will be a recurring theme throughout the book.

The book is divided into four parts. The theme runs through all of them, but the parts are quite different. Part I is ideas and history; Part II discusses relevant topics in programming with R; Part III treats object-oriented programming in some depth; and Part IV largely consists of a new approach to interfaces from R. An uneven landscape for sure, but one I found unavoidable if the key ideas were to be included in a single volume of reasonable size. I will try to provide trail guides and signposts, here and in the introductory page to each part.

Working backwards, the last two parts deal with two techniques especially relevant for serious projects: extending R's data structures through object-oriented programming (OOP) and using interfaces from R to other software. OOP (Part III, page 129) is the key technique for coping with complexity. By capturing the essence of new data structures in classes and methods, the user's R expressions remain simple and familiar, but are interpreted in the context of the new structure.

Interfaces (Part IV, page 241) are an integral part of the R software. As with OOP, they allow major new computational capabilities to be incorporated in a simple form for the user. Interface programming is also an important tool in adapting R to big data or intensive computational requirements. Part IV includes a new structure for interfaces, applicable to a variety of languages. The goal is to provide application package implementers with a simple and consistent view and to make the interface to other software essentially transparent for the users of the R package.

Part II of the book presents the apparatus that supports these and other extensions (page 59). The basic building blocks are functions. *Writing* functions is an elementary part of R. *Designing* functions that provide effective extensions of capabilities in R depends on understanding how functions and objects work, topics we emphasize in this part. We then go on to the next step up in scale for extensions, the R package, which is the most important medium for communicating new software in R.

These chapters do not attempt to give a complete guide to programming with R. I assume readers come to this book from a desire to extend R, which implies some experience with using it and some ideas for useful computations. There are a number of resources for guidance on many different aspects of programming with R, both books and a variety of online contributions. In particular, some topics will be filled in by reference to my own previous book [11] (not claiming superiority of this to alternatives, of course).

Part I is about understanding the concepts and background for the later parts. It presents the three principles and related ideas, looks at the evolution leading to current R in this context and outlines how R works.

## A little example

Extending R can contribute in many ways, often starting quite modestly.

Let me share an example, from one of my own interests, bird watching. It illustrates the relevance of extensions to R and in particular of the themes emphasized in Parts II through IV.

After a bird walk, it's good to record the experience to share and for future reference, noting what species were seen, interesting behavior or particularly beautiful birds. But in a two hour walk we may see 30, 40 or even more species. Few of us can remember all that. Making notes during the walk is distracting, plus you still need to organize them later.

My teacher about all things to do with birds, John Debell, showed me a better way: a spreadsheet with rows for all the species known to have been sighted in this area. That's several hundred rows, but recognition memory is stronger than recall memory, so one can usually mark the birds seen, after the walk.

Bird watchers are an idiosyncratic bunch, though, and his spreadsheet wasn't quite what I wanted. Also, cleaning up the empty rows and putting the result into a nice report for keeping and sharing was more spreadsheet programming than my meager experience could manage.

R, of course, was the obvious solution. Not that R is a spreadsheet program, nor is it built specifically to produce nice reports. Rather, the *extensions* to R include several contributions that made my task much easier and more pleasant. The starting point was an Excel spreadsheet form with species names and an indication of frequency seen (to help decide uncertain sightings). I would fill in a copy of this from today's walk.

The XLConnect package provides an interface to Excel, representing workbooks and sheets as R objects from which data can be read as an R data frame. Simple R expressions clean up the table, at which point there are many possible interfaces to presentation software, from reports to web displays. The function to construct an R object from the spreadsheet is on page 255 as an example of the Excel interface. Task accomplished, with very little pain.

Extensions to R were the key. First, the organization of software through functional computing and R packages such as XLConnect that makes such extensions available (Part II). Second, the ideas of object-oriented programming: XLConnect uses objects that are proxies for workbooks in Excel (Part III). Third, interfaces themselves (Part IV). Not only is XLConnect an interface itself, but it in turn builds on an existing interface, to Java.

As casual as the example was in its motivation, it also illustrates the benefits of programming in R to support unanticipated *future* extensions. My initial setup was for local use. But what about bird watching on a trip, in another area? One needs to replace the list of species with a check list for the new area. These exist (for example, at `avibase.bsc-eoc.org`) but need to be inserted into the existing form. R with the XLConnect extension makes this easy.

And John Debell has an extensive series of past records. Perhaps some interesting data analysis in the future?

## The Cover

Within the R community, contributors and interested users have joined together to form R User Groups, holding seminars and discussions and gathering to share experiences. The groups are a vibrant part of the R community, very much in the spirit of this book I hope.

The maps on the cover show the known user groups as of 10 years after the first release of R and 5 years after that. The increase of nearly a factor of 10 in the number of groups, and their spread throughout the world, illustrate graphically the growth in the R community. And specifically, the contributions of the thousands of packages and other tools that extended R during this period.

The R user group data were kindly provided by Joseph Rickert of Microsoft. For details and an up-to-date list, see

> `http://msdsug.microsoft.com/find-user-group/`

Also, `http://r-users-group.meetup.com/` reports R-related groups using `meetup.com`: over 200 groups with about 100,000 members as of January, 2016.

## Packages

The interface, tools and examples packages cited in the book are available from `https://github.com/johnmchambers`, including all the XR interfaces, XRtools and the examples presented for the interfaces to Python, Julia and C++. Some of the packages may be available from repositories as well; the individual package web pages on the GitHub sites will give the up-to-date information.

## Acknowledgements

To adequately thank all those who have helped me with their ideas, suggestions and opinions over the years would simply be impossible, even if my memory for details was much better than it unfortunately is.

Starting with the specific, thanks to Michael Lawrence, Dirk Eddelbuetel, Julie Josse and two of the publisher's reviewers for detailed comments on the draft versions.

Very special thanks to Balasubramanian Narasimhan for many valuable discussions and for the Statistics 290 course at Stanford, where I have had the opportunity to try out ideas and expound opinions over the years. Thanks also to the students and teaching assistants of that course for their questions and comments.

R, and S before it, have always been collegial projects developed by members of a community for that community. As emphasized throughout this book, the community involvement is the single most important strength of the project. So the debt I feel most strongly, and the thanks due, are to that community. Most directly today to my colleagues in R core, and to the many others in the community who support the process that makes R and its extensions available to the world. Looking farther back, to those who contributed to the ideas and software in S. From the beginning, the S project was an exciting joint project at Bell Labs, exploring ideas and their implementation for data analysis and research. Chapter 2 tries to convey some of the history. Although S was not open-source software, it began the growth of a community of users and contributors that later blossomed into the R community.

And, as in our previous seven books together, it has been a pleasure to work with my editor John Kimmel. A happy journey over 30+ years; may it continue.

# Part I
# Understanding R

This book is about extending R.

Doing so in a way that fits in effectively with the current software and uses it to best advantage requires understanding what's being extended. R is unusual in several respects, reflecting both its goals and the evolution leading to the current software. Its nature can be understood better by appreciating some fundamental characteristics, described in Chapter 1 as three principles.

The contents and features of R also reflect the evolution of the software (Chapter 2), including the S language on which R is modelled, the original creation of R and finally the infrastructure that supports the current version. Two general themes in computing that evolved over the same period—functional programming and object-oriented programming—influenced the evolution and are important for future extensions.

With the first two chapters as background, Chapter 3 looks at what actually happens when R is used.

# Chapter 1

# Objects, Functions and Interfaces

The purpose of this book is to encourage and assist extensions to the R software. Each such extension is something of value, both to the authors and to the community at large.

In a sense, nearly everyone who does any serious work with R is likely to be extending it. R is designed to make this easy, in the simplest case by defining one or more new functions. Any introductory book or online guide will introduce this procedure, and there's very little to it.

That may be as far as you need to explore, particularly when you are starting out. This book, however, takes "extending" in a stronger sense: Fundamentally, we will go on to the stage at which your contribution starts to have greater scope and potential value, both to you and to others.

My assertion is that more ambitious extensions benefit from understanding some of the underlying design of R. Three principles will be introduced that encapsulate much of that design. The principles reflect the fact that R, and the S language that preceded it, were created to bring the best software for computing with data to users in an immediate and effective form.

Understanding R can guide you to creating extensions that fit well with its essential capabilities. The creative process will be easier and more rewarding and the results are likely to be of greater value.

# 1.1 Three Principles

Throughout this book, three fundamental principles for understanding R will be our guides to extending R. They have been central to the software, since the version of S leading to R and even before.

Here they are in their simplest form, with the logos that will mark references to them throughout the book.

EVERYTHING THAT EXISTS IN R IS AN OBJECT. OBJECT

EVERYTHING THAT HAPPENS IN R IS A FUNCTION CALL. FUNCTION

INTERFACES TO OTHER SOFTWARE ARE PART OF R. INTERFACE

The details and implications of each of the principles will be recurring themes in discussing all aspects of extending R.

R has a uniform internal structure for representing all objects. The evaluation process keys off that structure, in a simple form that is essentially composed of function calls, with objects as arguments and an object as the value.

R and the S software that preceded it have always explicitly included the concept that one was building on a base of computational tools. The implementation of R remains consistent with the OBJECT and FUNCTION principles, relying on the ability of functions to communicate with and exchange objects with software in other languages. Initially, this provided the substantial toolkit of computational methods required to make the new system useful. The INTERFACE principle continues to be a key strategy for serious extensions of R.

The first two principles are related to two corresponding general paradigms in computing: *object-oriented programming* (OOP) and *functional programming*. These have proven to be among the most productive and popular paradigms for programming, over a long period of time. Many areas of application and many programming languages and systems have been based on or influenced by one or the other, especially by OOP.

Object-oriented programming is particularly valuable in managing complexity. Structuring data and computations through OOP allows software to evolve and deal with specialized requirements while retaining clarity and readability.

The techniques of functional programming support quality, reliable software, demonstrating the intent of the software and ideally supporting that with demonstrably valid computation. When the quality of the software is in question on theoretical grounds or in important applications, such support is valuable. It also gives useful guarantees, for example, that multiple function calls can be evaluated in parallel.

Statistical computing, a focus of our interest here, has a natural relation to both paradigms, but we've followed our own path. The software that has evolved into modern versions of R is distinctive and perhaps unique in that it has combined aspects of functional programming with several versions of OOP. Sections 1.5 and 1.6 introduce functional and object-oriented programming, at a very general level but in the forms we will explore through the rest of the book.

In both cases, the general ideas are highly relevant, but R takes a different approach to that found in other languages. Understanding how and why, in my experience, helps to avoid confusion, particularly if your own programming background is primarily in other languages.

I will refer to R's approach as *functional, object-based computing*. Functional computing and object-based computing are natural bases for functional programming and object-oriented programming, but looser. As implemented in R, neither imposes the discipline of the formal programming paradigms.

The approach in R reflects in part its own evolution. Chapter 2 discusses that evolution, with the goal of justifying the approach, or at least of explaining it.

With the principles and evolution as background, Chapter 3 gives essentials of how this all works in practice. Parts II to IV proceed to details of extending R, generally and then through object-oriented programming and interfaces.

## 1.2 Everything is an Object

Initial involvement with R naturally focuses on the specialized data and results of immediate interest. The data to be analyzed or processed may be specialized, for example, to that derived from spreadsheets.

```
> w <- read.csv("./myWeather.csv")
> class(w)
[1] "data.frame"
> str(w[,"TemperatureF"])
 num [1:92] 72.7 72.7 72.7 72.5 72.5 72.5 72.3 72.3 72.3 72.3...
```

The first programming tasks may be to produce equally specialized results: "What was the median temperature?". Easy to do in R.

The next stage is often to realize that the same questions are being asked repeatedly, maybe with some variations. Perhaps the results are now being used in a report or shown on a web page. This is what we'll call "programming in the small". R is very oriented to this too, the key step being to define a function, sooner rather than later (page 62).

Such specialized activities may still be unconcerned about a general view of the objects used. As a project gets more serious about extensions and new ideas, however, a general understanding will help to make effective choices.

Common forms of data in the usual sense of the term—vectors, matrices, data frames and everything else—are instances of the R object. Objects are created, manipulated and managed dynamically by the R process. At the basic level, all objects are equal: they all have certain intrinsic properties, which the user can query and (usually) set but which R itself will manage. This uniformity is built into the the standard R implementation, in the sense that all objects are instances of one particular type of structure at the C level.

But the ⌐OBJECT¬ principle does indeed mean *everything* that exists. The functions used or created in programming are objects.

```
> class(read.csv)
[1] "function"
```

So are the expressions used to call those functions or operate on objects. So are the objects defining the contents of an R package, at least when the package is loaded into the R process. At both the function and package levels, the ⌐OBJECT¬ principle is important when we want to extend what R can do.

Packages for programming, software development and testing have been developed in and for R by using function and language objects.

```
> ### All the global references in the function
> codetools::findGlobals(read.csv)
[1] "read.table"
```

A number of such packages are included with R (codetools) or can be found in repositories (e.g., devtools on CRAN).

Other tools deal with the packages themselves as objects. Once a package has been loaded into the R process, the corresponding *namespace* is an object (an "environment") containing all the objects from that package:

```
> ns <- asNamespace("codetools")
> str(ns)
<environment: namespace:codetools>
> str(ns$findGlobals)
function (fun, merge = TRUE)
```

Tools that aid in developing packages use namespaces and similar objects to analyze what software the package contains or depends on. We'll look at a number of examples in Chapter 7.

Distinctions among objects will be based on properties they have (their class, dimensions, names of the elements, $\cdots$). All the distinctions are *dynamic*; that is, they are queried dynamically in the R process, from the object itself.

R differs from some other languages that look rather similar in their treatment of objects, but which have primitive data that is not part of the general object system. In particular, single numbers and single character strings in R are not distinct from numeric or character vectors that only contain one element. Typing a single number causes the R process to create a vector of length 1, only differentiated dynamically from a vector containing thousands of numbers.

```
> 2
[1] 2
> sqrt(2)
[1] 1.414214
```

The absence of scalars is related to the OBJECT and INTERFACE principles. The original design of S represented objects in a form designed to communicate arrays to Fortran, and did not include scalars as such objects. As we will note in Chapter 2, there was no intent to include basic computations in the language, so the absence of scalars didn't bother us much.

One could restate the OBJECT principle as: R is an object-based system. Chapter 5 discusses object-based computing in more detail. The term is deliberately different from object-*oriented* programming, which adds some more structure to deal with classes and methods. That paradigm is important for extending R as well. Section 1.6, page 14 presents the connection in general, and Chapters 10 and 11 the two implementations in R.

# 1.3 Everything is a Function Call

The user of R interacts through a computing environment in which expressions are parsed and evaluated. The language in which these expressions are written looks like a number of interactive languages for quantitative computing (Python, Julia and others) and is similar to the procedural expressions in more explicitly "programming" languages (Java, C++). All these languages have function calls as part of their syntax but also assignments, loops and conditional expressions.

The family resemblance among the languages is not coincidental. Successful early languages, particularly C, established a syntactic style for "scientific computing", as it was called then. The languages mentioned above and others customized C-like syntax in various ways but preserved the basic form. Other syntactic styles have come and, for the most part, gone. (One alternative that *has* persisted, at least for a minority of users, is the purely functional form as in Lisp, which will be useful to understand function calls in R.)

In the case of R the C-style syntax is somewhat misleading. The diversity of expressions is largely an illusion, and disappears once the parser has worked over the input. True to the OBJECT principle, the parser's job is simply to convert the text into an object that represents R's interpretation of what you typed.

Asking "What happens in R?" is asking what the R evaluator will do given any possible object. There are only three cases: a constant such as the result of parsing "1.77"; a name, from parsing a syntactic name such as sqrt or any string quoted by back-apostrophes; and a function call. Anything that does not parse as a constant or name is represented in R by an object with the internal type of "language". Effectively this is a function call with the first element usually the name of the function.

The expression sqrt(pi) is parsed as a call to the function named sqrt, as you would expect. But operators, loops and other syntactic structures are evaluated by function calls as well: pi/2 is a call to `/` and { sqrt(pi); pi/2 } is a call to `{`. The whole braced expression could be represented functionally:

`{`(sqrt(pi), `/`(pi, 2))

Even more suggestive of the actual function-call objects would be notation in the style of the Lisp language, where the function name and all arguments are written inside parentheses, separated by commas:

(`{`, (sqrt, pi), (`/`, pi, 2))

As this notation suggests, a function call is a list-like object with the name of the function as the first element, followed by the expressions for the arguments. Section 3.2 looks into how the call and function objects are used in evaluation.

This form of computing, which we will call *functional computing*, means that the evaluation of any expression in R can be viewed as a "pipeline" in which the values returned by function calls become arguments to outer function calls.

The OBJECT and FUNCTION principles together can be summarized as a characterization of R:

> R IS A FUNCTIONAL, OBJECT-BASED LANGUAGE.

The principles can help in understanding specific computations in R, as illustrated by the following example.

Suppose **signature** is an optional argument to a function we are writing. If supplied, we expect a character vector. The default value is NULL. If we do have **signature**, we want to construct a single string, **tag**, by concatenating the elements of the vector. (This is how methods are stored in tables.) If the argument is missing or `NULL`, we won't construct the tag but will do a different computation. A compact way to implement this strategy in R is:

```
tag <- if(is.character(signature))
        paste(signature, collapse = "#")
```

There's a slight subtlety to this expression, which our two principles help clarify: What happens if `signature` is *not* a character vector?

First, the expression is an assignment; by [OBJECT] therefore, after it finishes `tag` *will be* an object. Second, by [FUNCTION], the right-hand side of the assignment is evaluated as a call, to the function `` `if` ``, just hidden a bit by the conventional notation for a conditional expression. A complete `if( ) ... else ...` expression calls the `` `if` `` function with three arguments. An expression without an `else` clause omits the third argument. The result when the condition is `FALSE` is then the value of the function call when the third argument is missing.

If we look up the documentation in `?if` we discover that the value of the expression will be `NULL` in this case. The expression shown is therefore a valid and compact way to define `tag`, if not perhaps ideally clear. Later on, the calling function should test `if.null(tag)` to branch correctly.

In some languages, a conditional expression does not behave like a function call, and the syntax used above might be illegal. The conditional expression would be required to have a form like:

```
if(is.character(signature))
    tag <- paste(signature, collapse = "#")
```

which could leave `tag` undefined unless there was an `else` clause or an assignment earlier in the computation.

Functional computing is sometimes loosely referred to as functional *programming*, but that term should really be reserved for a stricter definition: In functional programming, the result of the computation is entirely the value returned, with no side effects, and is entirely determined by the arguments, with no dependence on other data. R and some other languages that support a functional style of computing do not enforce this strict form, but R encourages it by its implementation of function calls and local assignments. Functional programming is an important concept for good software development in R: Section 1.5 relates the general paradigm to computations in R.

# 1.4 Interfaces are Part of R

The whole of Part IV of the book is devoted to interfaces for extending R. They come in many forms and can be crucial to incorporating exciting new techniques.

But, regardless of what you do in a standard R session, you will still be using interfaces. For the R evaluator, everything that happens is a function call. In particular, no pieces of code will be compiled into machine code or interpreted directly at that level. Those basic computations do take place, but in functions that provide an internal interface to the machine-code level. There are a number of such internal interfaces in R, which we will look at in Section 5.4. The only ones useful for extending R are those that call a compiled subroutine dynamically linked to the R process. The base package sometimes uses a more low-level interface to C routines, which short-cuts general function call evaluation. Although this complicates the analysis somewhat the essential point remains: all the basic computations that are not themselves programmed in R are done through a subroutine interface.

In Chapter 2 we will see the essential role of interfaces in the history of R, back to the first version of S.

In the context of extending R, the most important reason for interfaces from R is to do something new. For a computation relevant to our goals, either an implementation already exists in another language or programming the computation in another language is the most attractive option. Part IV of the book will present approaches to interfaces for a variety of languages, including interfaces to subroutines but also interfaces to evaluators in languages such as Python and Julia and to languages for data management and organization.

Efficiency is another motivation for doing computations outside R, sometimes for good reasons and sometimes unnecessarily.

The absence of machine-code level computations may be of concern if we believe computational efficiency is a limitation. Everything is a function call: R evaluates the expression

```
pi/2
```

by calling the function `` `/` ``:

```
> `/`
function (e1, e2)  .Primitive("/")
```

This is one of the primitive interfaces; basically, the evaluator looks up the function definition, finds that it is a special kind of object and uses that object to call one of a collection of known subroutines, immediately. The computation is a good deal simpler and faster than the standard R function call, relying on the function being found in a table of built-in primitives. The mechanism is fundamentally not extensible.

More efficient as it is, this is still an R function call. Let's look for example at a line from a silly R implementation of a convolve() function (Section 5.5, page 91):

```
z[[ij]] <- z[[ij]] + xi * y[[j]]
```

This expression calls four functions: `` `+` `` and `` `*` `` and two versions of the operator `` `[[` ``, an extraction version called twice on the right and a replacement version for the call on the left. As you might expect, all of these are primitive functions, but even so there are 5 R function calls to compute one number each time the expression is evaluated, which will clearly be in a loop. Compared to machine-level fetches, stores and arithmetic this is a lot of computing. That is why one is inclined to call this a silly implementation.

So what? I do computations nearly as silly frequently in R when I want some interesting result that can be easily programmed. Nearly always, the computation completes before I even begin to wonder at how inefficient it is. Somewhat more relevantly, colleagues have shown me a number of their implementations of ideas in statistical methodology that would have seemed prohibitively inefficient but which returned the answer well within their personal tolerances.

There are, on the other hand, increasingly important applications where the volume of data and/or the computations to be done make computing time a critical issue. It is possible, but not guaranteed, that the speedup from implementation in another language may help substantially. Section 5.5 will have more to say.

# 1.5  Functional Programming

The basic principles of functional programming can be summarized as follows.

1. Computation consists of the evaluation of function calls; programming therefore consists of defining *functions*.

2. For any valid function definition in the language, a call to that function returns a unique value corresponding to each valid set of arguments, and *only* dependent on the arguments.

3. The effect of a function call is solely to return a value; it changes nothing that would alter any later computation.

The first point implies R's ⃞FUNCTION⃞ principle: everything that happens is a function call.

The implication of the second point is that functions in the programming language, like functions in mathematics, are mappings from the allowed set of arguments to some range of output values. The returned value should not depend on the "state" of the evaluator when the function call is evaluated, sometimes expressed as the requirement that the computation be *stateless*. The third point says that the function call should not change the state; that is, it should have no *side-effects*.

The functional programming paradigm applies naturally to statistical models. An elementary function for fitting a linear regression model might have the form:

```
lsfit(X, y)
```

In its geometric formulation, for example, linear regression is defined in functional terms: "The least-squares linear model for $y$ on $X$ is the projection of $y$ on the space spanned by the columns of $X$." The model is a *function* of the data. If we fit the model again tomorrow the result should be the same, ideally even in a different computing environment.

The result should not depend on tinkering with other quantities beforehand. Nor does computing the regression alter anything else in the computing environment.

An idealized functional programming unit is then a self-contained definition, from which one could infer the function's value for any specified arguments—a *definition* as opposed to procedural instructions for how to calculate the result. Here, for example, is a definition of the `factorial()` function for integer scalars in the Haskell language:

```
factorial x = if x > 0 then x * factorial (x-1) else 1
```

There is also the requirement, not included in the definition itself, that `x` be an integer scalar. In practice, there must be some procedure for expanding the definition, given actual arguments. This expansion has much of the flavor of R's evaluator, which processes a function call by evaluating the body of the function in an environment containing the arguments as objects (we'll examine the details in Section 3.4).

In simple examples, such evaluation will be essentially identical to a functional program. The definition of `factorial` in Haskell could be an R function, aside from differences in syntax:

```
factorial <- function(x) {
    if(x > 1)
        x * Recall(x-1)
    else
        1
}
```

In R, using `Recall()` instead of `factorial()` ensures that the recursion calls the correct function, as an object—the $\boxed{\text{OBJECT}}$ principle again.

Over the decades since functional programming was introduced, in the 1970s, there have been a number of languages built explicitly around the paradigm. The functional computing paradigm has also been incorporated within languages that do not strictly follow functional programming, as we did with S. That approach continues to be pursued, for example, in a functional extension to JavaScript, in order to "overcome JavaScript language oddities and unsafe features" (in the publisher's blurb for [17]). The Julia language follows the functional computing paradigm but without preventing side effects.

The first requirement raises the question of what a function definition *is*, and how such a definition can incorporate other functions. No practical language can require that all function definitions be literally self-contained. That works for such elementary examples as `factorial()` but serious programming relies on libraries of functions or the equivalent. Pure functional languages treat external definitions as "short forms" for including the relevant code in place of the calls; essentially what would have been called macro definitions. Modern functional languages tend to be more sophisticated than old-style macro systems; for example, Haskell uses some powerful notions of lazy evaluation of both function calls and objects.

Really achieving pure functionality is anything but trivial, particularly when embedded in a more general language. Functional programming embedded in languages with non-functional aspects is generally partial and dependent on programmers choosing the functional portions of the language.

In addition, most important results for quantitative applications, such as fitting models, cannot be computed with total accuracy. At the least, the "state" of the computation enters in the sense of the machine accuracy. In R, some computational building blocks must be assumed; for example, the basic computations for numerical linear algebra. These can differ between installations of R.

Nevertheless, a functional perspective can been very fruitful in thinking about important computations. The FUNCTION principle ensures that R programming focuses on defining functions. Particularly early on in a project, a strength of the language cited by many users is the ease with which one can create functions corresponding to computations of interest. Typical interactive programming with R involves writing some functions, then calling those functions with the data of interest as arguments. If the results aren't as desired or one thinks of a new feature, the functions are rewritten and the process continues.

Whether the functions obey functional *programming* requirements is a different question. The short answer is that R can support functional programming, but it's not guaranteed or automatic. As a desirable goal and a way to understand the R functions being designed, however, it is valuable. Section 5.1 examines the criteria for functional programming as they apply to extending R.

# 1.6    Object-Oriented Programming

Object-oriented programming in the form it has developed over a long period of time rests on a set of concepts. Stated briefly:

1. Data should be organized as *objects* for computation, and objects should be structured to suit the goals of our computations.

2. For this, the key programming tool is a *class* definition saying that objects belonging to this class share structure defined by *properties* they all have, with the properties being themselves objects of some specified class.

3. A class can *inherit* from (contain) a simpler superclass, such that an object of this class is also an object of the superclass.

4. In order to compute with objects, we can define *methods* that are only used when objects are of certain classes.

Many programming languages reflect these ideas, either from the start or by adding them to an existing language, in some cases without formalizing the concepts completely.

Part III of the book will present the implementations that have been added to R for object-oriented programming. How does R itself relate to OOP, in particular in terms of the OBJECT and FUNCTION principles?

The first point in the list above is clearly the OBJECT principle. Objects are self-describing; that is, the object itself contains the information needed to interpret its structure and contents. Different types of objects are built into R and mechanisms exist to incorporate structural information (in particular, R *attributes*).

This object-*based* approach is an important step in making statistical computing more accesible to users, but it does not imply any of the other points in the list. Chapter 2 will show some of the changes along the way to the current R language that introduced the remaining aspects of object-oriented programming in stages, as optional extensions of the basic language.

In discussing OOP in R, there is a crucial distinction to be made. The last point in our list above says that methods will be the mechanism for implementing class-dependent computations but says nothing specific about what a method is or how it would be used. In fact, there are two quite distinct approaches. Both are available in R and will be described in detail in Part III. The two versions will be referred to as *encapsulated* and *functional* object-oriented programming. The distinction is whether object-oriented programming is integrated into a functional computing framework, where the FUNCTION principle applies.

Traditionally, most languages adopting the OOP paradigm either began with objects and classes as a central motivation (SIMULA, Java) or added the OOP paradigm to an existing language (C++, Python) as a non-functional concept. In this form, methods were associated with classes, essentially as callable properties of the objects. The language would then include syntax to call or "invoke" a method on a particular object, most often using the infix operator ".". The class definition then *encapsulates* all the software for the class. Methods are part of the class definition.

Given the $\boxed{\text{FUNCTION}}$ principle in R, the natural role of methods as an extension of functional computing corresponds to the intuitive meaning of "method"—a way to compute the value of a function call. For a particular call to the function, a computational method is chosen by matching the actual arguments as objects to the method that most closely corresponds to those objects. Specifically, the methods will be defined with a *signature* that says what class(es) for the argument(s) are assumed. Methods in this situation belong to functions, not to classes; the functions are *generic*—we expect them to do something sensible, but different, for different classes of arguments.

In the simplest and most common case, referred to as a standard generic function in R, the function defines the formal arguments but otherwise consists of nothing but a table of the corresponding methods plus a command to select the method in the table that matches the classes of the arguments. The call to the generic is evaluated as a call to the selected method.

We will refer to this form of object-oriented programming as *functional* OOP and to the form in which methods are part of the class definition as *encapsulated* OOP.

Encapsulated OOP is the form most widely associated with the term object-oriented programming. It is available in R and the natural choice when the object is one expected to change but persist—a *mutable* object in OOP terminology. Such objects arise in many contexts from certain kinds of models to web- or graphics-based tools. They will also be natural when we come to discuss interfaces to other software using encapsulated OOP.

The alternative paradigm, functional object-oriented programming, is used in R and a few other languages. The most relevant other implementation for current and future projects is the Julia language (www.julialang.org). Functions are specialized as methods in Julia much as in R, according to Julia types. We will discuss an interface to Julia in Chapter 15. The Dylan language [32] also has a version of functional OOP; the language has been the subject of computer science research. It takes a stricter, more formal view than R or Julia.

Sections 2.5 and 2.6 discuss encapsulated and functional OOP, in the evolution of R. Part III of the book is about object-oriented programming for extending R.

# Chapter 2

# Evolution

This chapter examines a path through the history of statistical software, in particular, software to support data analysis and statistics research. The path follows the evolution that would result in the current R language and its extensions.

The direct evolutionary path begins with the first version of S, described as "an interactive environment for data analysis and graphics" [3]. S was created by statisticians with an interest in computing, members of the the statistics research community at Bell Labs. Statistics research at Bell Labs involved a combination of research into methodology and challenging, non-routine data analysis. S was designed for the computing needs of these activities.

S evolved through several versions and was used increasingly within its original corporate home and then outside via a licensing arrangement. Statistics departments at universities in particular became users of S and contributors of software to extend it. Some twenty years after the first plans for S, the R language was introduced as "A language for data analysis and graphics". An international group of statisticians with an interest in computing then developed R into a free, open-source system incorporating the facilities of S within the design of the new language, releasing version 1.0.0 of R in the year 2000.

Since then, R has grown, evolved and been extended. The evolution through this path from the first S to current R will be examined in various sections of the chapter. To understand the evolution, we also need to understand some of the external influences and software crucial to it. Section 2.1 considers the relevant evolution of the essential software to which the original S provided an interface. Sections 2.2 to 2.4 follow the evolution of S and of R. Section 2.5 then detours to examine the ideas of object-oriented programming, an important influence on later stages and the topic of Part III of the book. Sections 2.6 and 2.7 review later evolution of S and R.

# 2.1   Computational Methods

The two decades preceding the first distribution of S in the 1980s saw substantial work in statistical computing. Statistical societies established sections and working groups on computing. Professional meetings devoted to statistical computing became regular events.

A number of software projects were carried through, several of which produced statistical systems that were used extensively, notably BMDP and SAS. None of the major systems were ancestors on the path leading to S and R, however.

The most heavily used "systems" or "packages" aimed to provide high-level statistical computations via command languages, usually organized by keywords corresponding to the analysis or data manipulation operation. Here is a linear regression from the BMDP manual [14], simplified and omitting a few lines:

```
/ PROBLEM   TITLE IS 'WERNER BLOOD CHEMISTRY DATA'.
/ INPUT     VARIABLES ARE 5.
            FORMAT IS '(A4, 4F4.1)'.
/ VARIABLE NAMES ARE ID, AGE, WEIGHT, URICACID, CHOLSTRL.

/ REGRESS   DEPENDENT IS CHOLSTRL.
            INDEPENDENT ARE AGE, WEIGHT, URICACID.
/ END
```

Along with suitable input data, this would provide a batch "job" to run. The result of the linear regression in this example would be some printed output summarizing the input problem and the computation performed.

The systems typically provided data processing and analysis as a service. The use of specialized sets of keyword phrases could be characterized as attempting to avoid a programming language, not to design one.

This approach was not suitable for statistics research or challenging data analysis. More relevant for the evolution of R was software written in a succession of programming languages to implement techniques for "scientific computing". Three aspects of the developments are particularly relevant for our story: the organization of data; the structure of the programming; and the resulting software libraries. (If you suspect a connection with the three principles of Chapter 1 you are entirely justified.) All three aspects evolved dramatically during the first few decades of the digital computer.

With computers and digital devices embedded in nearly everything today, it takes some effort to appreciate the conceptual impact of the basic ideas behind the computer. The term "stored program computer" captures the most central innovation: Coded instructions for operations (a *program*) and data on which the instructions operated would both be stored on a memory device. The machine would start by loading and executing an instruction from one location and would then proceed (usually) to the next instruction.

Compared to earlier devices, the scope for computations was revolutionary. For example, I "programmed" the IBM Sequence-Controlled Calculator by coding instructions onto an external wiring board. Data input and output were on punched cards. The columns of a table were represented by fields on successive cards in the input. The wired program could instruct the calculator to perform one of a limited number of numeric operations, and then to punch the result onto the card as output. Thus, computing a linear operation on one or two columns in a table was possible, and essentially the conceptual limit of the device.

Figure 2.1: The IBM 650 Computer, mid-1950s.

In contrast, the flexibility of a stored program and the ability to examine and potentially alter data and instructions in a general way offered conceptual possibilities that we still have not exhausted. Over the period up to the early 1970s, key ideas were introduced about organizing data, about the structure of the programs and about the numerical and other techniques to compute relevant results. The ideas can be seen in the software we use today, notably in R.

The IBM 650 (Figure 2.1) was probably the first digital computer widely sold and used (and the computer on which I did my first programming, around 1960).

It belonged to the last generation of pre-transistor computers, when these were regarded as inheritors of the electro-mechanical machines like the Calculator—the "business machines" that had made IBM successful over the previous decades. Computers of this era cost a great deal of money, occupied much of a room and generated considerable heat. (One of the perks of a summer job working with a computer was that it might be in the only air-conditioned room in the lab.)

## Data structure

When programmable digital computers were first applied widely to scientific calculations in the 1950s, the data they contained was addressed by offsets into physical media in fixed units, such as bytes or words.

The data storage on the 650 used a rotating magnetic drum with a few thousand words of storage. (The gentleman in Figure 2.1 is pointing to it: It occupies the whole lower third of the cabinet.) Words could be read or written only when the read or write head was over the particular location, so the physical placement of one's data was clearly important to computational efficiency!

The supporting software for the 650 was IBM's symbolic assembler program SOAP [21]. This had declarations for individual words or blocks of storage in the user's program. The assembler attempted to locate instructions and data on the drum for efficient computation (the "O" in the name stood for "Optimizing"). The focus remained on the storage as hardware and not on the data represented. In any case, the physical limitations would have left little room for thinking beyond the immediate calculation.

Even at this stage, some abstraction of data was possible, opening the way for expressing important computations for quantitative applications. In SOAP, one could associate a symbol (one character only) with a region on the drum:

```
REG x 1000 1099
REG y 1100 1199
```

This would reserve two blocks of 100 words labelled x and y. Instructions in the computer were suitable for iterative computations on the corresponding words in the two regions, such as numerical combinations of values from each (just like the Sequence-Controlled Calculator, but no need to punch cards).

Here already is the basic notion of vectorized computations: performing the same calculation on all elements of two similarly structured vectors. The power of the stored program concept was that the calculation could be elaborated as needed, beyond simple operations that could be hard-wired for earlier devices. Basic operations in numerical linear algebra, for example, were a natural application.

As the computing capacity of the hardware expanded by many orders of magnitude, the software used to implement computations evolved also. By 1960 "higher-level" languages such as **Fortran** had already taken important steps forward from the mechanistic view of data, although that view can still be seen in the background. **Fortran** allowed programmers to abstract a reference to a location or block in memory as a symbol in the user's program, to be converted into physical terms by compiling and running the program. The content of the references evolved also, into the first *data types* describing the data contained, including integer or real (floating point) numbers, characters or logical values:

```
real x(100), y(100)
```

One still had to program in terms of single items or contiguous blocks of a specified size, but now the references to these included the type of data stored, and physical addresses were abstracted away.

In scientific computing particularly, a key change was to think of a contiguous sequence of numerical values—the **Fortran** *array*—as opposed to the actual memory regions where that data would be stored. Indexing the elements of the array by an integer variable was a natural mapping from the mathematical notation of subscripting elements of a vector. By a natural extension, the sequence of values could also be interpreted as a two-way array, by convention stored by columns. Pairs of indices again mimicked subscripted elements of a matrix. Higher-way arrays followed in the same format.

Now the concepts had evolved from physical storage to data, at least for numeric vectors and arrays. For statistical applications, this marked a watershed: The notion of observational data as a collection of $n$ values on $p$ variables could be mapped, with some approximations, to such structures.

## Program structure

As the treatment of data evolved from a direct relation to physical memory to include simple but powerful abstractions, similarly the treatment of the instructions for programming computers evolved from explicitly coding the machine's instruction set to a degree of abstraction, while still reflecting the hardware capabilities. In addition, the program as a whole took on important structure, an essential step in being able to share software and to extend the work of others.

As with data so with programming, the initial view was based on the hardware of the machine. Initially, programs had no internal structure. A program was defined by the contents loaded into locations in the memory—the "instructions". In the **SOAP** assembler for the IBM 650, the "S" stood for "Symbolic" but that referred only to the use of names for the numeric instruction codes, for locations in the program itself and for scalar or regional data locations. The computer would normally execute an instruction and go on to the next; instructions could cause it to branch to a different instruction unconditionally or on testing some hardware condition (typically in response to counting down a register from a preset length). The details naturally differed from one machine to another.

A program needed to be complete in itself. It would start, usually read in some input data for the particular run, carry out the complete computation including the generation of results (for example, a punched set of cards), and then halt or go into a loop waiting for human intervention. Figure 2.2 is a listing for a program in **SOAP** to compute a quadratic transformation of some data, read from and written to punched cards. The program itself would be key-punched from the handwritten listing.

**650 SOAP II CODING FORM**  $F_{(x)} = Ax^2 + Bx + C = (Ax + B)_{x + C}$

| LOCATION | OPERATION CODE | DATA ADDRESS | T A G | INSTRUCTION ADDRESS | T A G | REMARKS | ACCUMULATOR UPPER 8003 | ACCUMULATOR LOWER 8002 | DISTRIBUTOR 8001 |
|---|---|---|---|---|---|---|---|---|---|
| EXAMPLE | | b CAL | C | ULATE | | F OF X | | | |
| | B LR | 1951 | | 1960 | | READ AREA | | | |
| | REG P | 0027 | | 0028 | | PUNCH AREA | | | |
| SETX | RAU | ONE | | STX | | SET X | 1 | | |
| STX | STU P | 0001 | | | | TO 1 | X | | |
| | MPY | A | | | | CALCULATE | | AX | |
| | ALO | B | | | | F | | AX+B | |
| | RAU | 8002 | | | | | AX+B | | |
| | MPY P | 0001 | | | | | | (AX+B)X | |
| | ALO | C | | | | | | (AX+B)X+C | |
| | STL P | 0002 | | | | | | | |
| | PCH P | 0001 | | | | PUNCH | | | |
| | RAU P | 0001 | | | | IS X MAX | X | | |
| | SUP X | MAX | | | | | X-100 | | |
| | NZU | | | 9999 | | | | | |
| | AUP | 01 | | STX | | STEP X | X+1 | | |
| ONE | | 00 | 0000 | | 0001 | CONSTANTS | | | |
| XMAX | | 00 | 0000 | | 0100 | | | | |
| 101 | | 00 | 0000 | | 0101 | | | | |

Figure 2.2: A program listing in **SOAP** to compute a quadratic function of some input data, from [21].

At this level, the computer offered the potential for complex calculations but the intellectual barriers were prohibitive for creating significant software and even more for sharing it in the scientific community. In response, the first generation of higher-level languages appeared.

The instructions written in these languages were intended to be more meaningful to the human and therefore easier to write and perhaps even to read. The instructions would be interpreted for a particular machine, but were intended to be themselves relatively independent of hardware details. For the most widely used languages, the interpretation took the form of translating ("compiling") the program into either machine instructions or into a suitable assembly language.

While intended to be understood by the programmer, the language also needed to be sufficiently specific and complete to generate most if not all the possible programs one needed to run, and in a form at least efficient enough to carry out the computation in practice. So began a tradeoff that continues today and is one of the main topics of the present book: between the ability to express ideas as software and the need to carry out challenging computational tasks.

As far as the structure of the program was concerned, the single most important concept introduced was the *procedure*. A procedure was a segment of code in the language with the property that it could be entered by "calling" it, using some identifier and usually some list of arguments. The procedure would do some computations; when those were complete, it would "return", causing execution to resume at the point after the call. To be useful, most procedures would need to communicate some result either by putting a simple result in some machine register or known location, or by using the arguments to modify data specified by the call. The language and procedure innovations could have been done independently—an assembly language can have procedures in principle—but in practice they arose together in various forms through the early 1960s.

Of the languages introduced then, the one most likely to be familiar to R users is Fortran, which was the language underlying the first versions of S and was used in many of the libraries of computing tools for statistical and numerical purposes.

There were a number of other proposals at the time, two of which are worth mentioning here. The Algol family of languages was intended for coding algorithmic computations, as its name suggests. Many important procedures were first published as Algol. The features of the language were also influential in the design of later languages.

A less well-remembered example was the APL language, introduced in a book in 1962 [24]. APL's idiosyncratic design led to compact implementation of operations on matrices and multi-way arrays. Its influence on S was not so much the language itself (although we did take over some of its approach to arrays) but rather the concept of a language that could be used interactively and could provide real-time insight to the user, for example by automatically printing the value of an expression, along with extensions by new functions.

## Algorithms and libraries

The evolution of data structures and procedures outlined above may not sound like great stuff today, but a major revolution in scientific computing was supported by languages such as Fortran and by improvements in hardware. Numerical techniques were programmed and incorporated in libraries of software, implementing much more effective and reliable methods than had existed before for key computations in statistics and other disciplines. Direct descendants of many of these are still in use, such as the software supporting many of the computations in R.

Throughout the 1960s and into the 1970s, many authors worked out explicit descriptions of computations. These included techniques for numerical, symbolic and other calculations of relevance to scientific applications. The descriptions might be written in "pseudo-code"; that is, in a more-or-less formally defined language that was not claimed to be directly executable. Increasingly, however, publications and informal write-ups included code in an actual programming language, "algorithms", borrowing a term for a set of instructions to humans for solving a problem. Algorithms were published in sections of journals, including those for numerical analysis and some of the statistics journals. When implemented in a language such as Fortran or Algol they provided *subroutines* of potential value to statistics and other applications.

Collections of such subroutines formed *libraries*, with a greater or lesser portability over different computer systems. Within such a collection, there was a motivation to make the algorithms more usable by structuring them around a consistent view of related computations, such as those for matrix operations or for random number generation.

We had such a library for statistical computing in Bell Labs research. It was the main resource for data analysis and research before and, in fact, after the initial development of S. Its organization greatly influenced the structure of S and therefore of R.

My 1977 book *Computational Methods for Data Analysis* [8] was in effect organized along the same lines. Its chapter headings give a relevant view of the structure and also highlight major computational advances during the period, by many authors.

**Data management and manipulation:** Algorithms were developed for efficient sorting, partial sorting and table lookup. We also provided storage management utilities that later supported management of objects in S. Notably absent at this time was effective linking of scientific software to database management systems.

**Numerical computations:** Methods were developed for various transformations of numeric vectors: approximations, including splines; spectral analysis of time series, including the fast Fourier transform; numerical integration.

**Linear models:** A major advance in this period came from numerical techniques based on matrix decompositions. Previously, old hand-calculation techniques (solving "normal equations") had been used. The basic linear model functions in R still use revised versions of the QR and other decompositions from this period. Singular-value and eigenvalue algorithms were also important for statistical applications.

**Nonlinear models:** Nonlinear optimization also saw major advances, including special techniques for nonlinear least squares.

**Random numbers:** Digital computers opened serious possibilities for the first time to use simulation or Monte-Carlo techniques. Simultaneously, the quality of the basic generation was greatly improved. A combination of this with approximation techniques extended the specialized techniques for the uniform distribution to a general capability to produce quantiles and simulated samples from a variety of distributions.

**Graphics:** Software for non-interactive plotting, such as scatter plots, had been developed; we adapted a collection of Fortran subroutines into a system, [2], that became the graphics interface for S and for base graphics in R. Interactive graphics came late in the period and had not yet been integrated into the general-purpose software.

To make a substantial subset of the computational methods available in practice, in one organization, was a substantial undertaking. Nor was there anything like the internet to facilitate software exchange. The appendix to my book was a table of software sources and how to get them: It's relevant that the majority of items existed as "listing" only. However, an important precedent was that many of these algorithms were described in publicly available documents and "free", more or less explicitly, for others to use.

With all these challenges, statisticians in the early 1970s with access to a good library of procedures, such as we had at Bell Labs, had a much improved set of tools for data analysis and research.

How to use the tools more effectively to try out new ideas became a concern.

## 2.2   The First Version of **S.**

Statistics research at Bell Labs was involved with computing from the first generation of computers. The first meetings to discuss an interactive statistical system at Bell Labs in which I participated occurred in early 1965. (In an unpublished memorandum following those meetings, John Tukey referred to "the peaceful collision between statistics and computing".) That system never came into existence.[1] Over the next decade, efforts were concentrated on computational methods.

By the mid-1970s, we had an extensive in-house subroutine library along the lines outlined in the previous section. The software supported some reasonably ambitious applications of statistics to AT&T data and also research into some "computationally intensive" methods, such as robust estimation. Although the size of the data and the speed of the computations were orders of magnitude less than possible today on a laptop, the contributions of the software to the organization were very real.

Our goal with S was not to replace all this software, but to build on top of it a language and environment where statisticians could interact directly, expressing their computations and the structure of their data more naturally. At the same time, we were a research organization: any solution had to be extensible to accommodate new research and applications.

Consider applying classic linear least-squares regression, as an example. Our Fortran library at Bell Labs had a subroutine to fit a linear least squares model. That subroutine in turn relied on some publicly available software to perform the underlying numerical linear algebra.

To call the subroutine, the programmer would write something like:

```
call lsfit(x, n, p, y, coef, resid)
```

There are four arrays in this computation: The model fits a vector y to a linear combination of rows of the matrix X and returns a description of the fitted model in vectors of coefficients and residuals. Type declarations in the language can signal that X, y, coef and resid are arrays of (floating point) numbers, and that X is a two-way array. The lsfit procedure would have the form:

```
subroutine lsfit(x, n, p, y, coef, resid)
real x(n, p), y(n), coef(p), resid(n)
integer n, p
```

---

[1] Primarily because Bell Labs dropped out of the Multics operating system project, for which it was intended. Another consequence of that decision was the creation of an alternative operating system at Bell Labs—Unix.

The program calling the subroutine is responsible for ensuring that all the arrays are allocated with the correct dimensions and type, and for communicating that information correctly to the lsfit() subroutine.

As noted above, database management software was not usually included in the scientific libraries. Programming an application included reading in and organizing the data, usually via some site-specific use of external media, magnetic tape in many cases.

Once the computations are done, then what? Insights from the analysis need more than just examining the two output arrays. One needs to organize the output and then do additional programming to summarize and explore. We had become particularly convinced of the value of graphical displays for data analysis. Each of these summaries or displays required more programming.

So the computations in the lsfit algorithm were quite good, just a slightly less refined version of the numerical methods used today in R. The problem concerning us was the process needed to translate ideas into code that incorporated that algorithm.

The Fortran main program that called the lsfit() subroutine required a substantial amount of code, and each new idea required re-executing that program with changes. Each analysis was a "job" in the terminology of the time. The necessary Fortran programming and other specialized instructions (for example, job control to mount external media) required a separate skill set (usually provided by a team member other than the principal statistics researchers), required considerable time and naturally was subject to possible errors.

In other words, much human effort and considerable time came between getting an insight from some computations and using that insight to try something new. In a substantial statistical project, there was a high cost in human effort and in calendar time.

Those of us involved in the planning that led to S were looking for an alternative, one in which the software did more and computations could be expressed more directly. Human efficiency would benefit, even at the cost of extra machine computation.

S was designed as *An Interactive Environment for Data Analysis and Graphics*, as the first book [3] was titled. Data was stored in a workspace and results displayed directly in response to programmed expressions.[2]

The first version of S to be distributed, beginning in the early 1980s, expressed a least squares regression as:

---

[2]Iverson's APL system [24] had pioneered this mode of computing and might itself have been used, except that interfacing to the existing Fortran library was essential and not part of the APL paradigm then.

```
fit <- reg(X, y)
```

Arguments X and y are now self-describing objects; the object, not the programmer, provides information such as dim(X) for the dimensions of the matrix. Model-fitting has become a functional computation: given arguments defining the desired model, the function reg() returns a self-describing object containing all the relevant information, with components for the model coefficients, residuals and other computational information. After the fit, a normal Q-Q plot to examine the residuals could be produced:

```
qqnorm(fit$resid)
```

This all looks very R-like, but how was a function defined? How would one extend S?

We are moving towards the ⟨FUNCTION⟩ and ⟨OBJECT⟩ principles. Data and fitted model are objects; the function itself is not yet quite an object.

New functions were not created as objects but were programmed in a separate language, the "interface language". The interface language looked a little like S but actually was parsed and transformed into an extended version of Fortran. An interface function for reg() might look like this:

```
FUNCTION reg (
    x/MATRIX/
    y/REAL/
    )
if(NROW(x) != LENGTH(y))
    FATAL(Number of observations in x and y must match)
STRUCTURE (
    coef/REAL, NCOL(x)/
    resid/LIKE(y)/
    )
call lsfit(x, NROW(x), NCOL(x), y, coef, resid)
RETURN(coef, resid)
END
```

The name of the function and of its arguments are specified, along with type declarations for the arguments. After some error checking and the allocation of the data for the output arrays, the algorithm is called, as it would be from a Fortran program.

The design reflects the ⟨INTERFACE⟩ principle in the original version of S: Our intention was not to replace Fortran by using the interface language to write low-level code. Instead, its use was to prepare arguments and structure results from interface calls to subroutines. The title of the book describing the interface language, *Extending the S System* [4], indicates that we saw such incorporation of new computational methods as the central mechanism for extensions.

Declarations were provided for a limited range of S data objects, both for the arguments to the function (`x` and `y`) and for new objects to be created (`coef` and `resid`).

These were true S objects, but interpreted and embedded in an extended version of Fortran. When the name of an S object appeared in an ordinary Fortran context (`x` in the call to `lsfit()`), the interface language substituted the Fortran-level data in that object (the numeric array in this case). Built-in functions accessed information about the S object (the number of rows in a matrix, for example).

The code could include Fortran declarations for scalars and arrays, as well as the standard iteration and control structures. In effect, the programmer had the full resources of Fortran plus a variety of extensions. The interface language source would be compiled and linked into a library file that would then be attached to an S session.

As the main programming mechanism for users of S, the interface language was a stumbling block. The initial learning barrier was high, particularly for those unaccustomed to programming in Fortran.

The version of S on which R was based (Section 2.6), replaced functions in the interface language with true S functions. However, the interface language had important advantages, worth considering today in dealing with challenging computations where efficiency matters.

1. Because the interface language did end up as Fortran, it could produce computationally efficient code with direct access to much existing software for serious numerical methods.

2. At the same time, having a customized language allowed the programming to look more like S (and with less primitive tools, one could have gone much farther in this direction).

3. Some non-S programming was helpful for efficiency or clarity (such as type declarations for arguments) and could have been useful independently of having a particular subroutine to call.

These features remain relevant today and have in fact resurfaced in some current approaches to interfaces from R. The closest general analogue is the Rcpp interface to subroutines through C++ (Chapter 16), which includes a number of R-like programming constructions, called "sugar" in Rcpp. Other approaches to interfaces when efficiency is a concern are discussed in Section 12.8.

Objects in the first version of S reflected their Fortran origin. When a Fortran array (a contiguous block of the same type of data elements) became self-describing and managed for the user by the language, it became the S *vector*, still the basis for most objects in R today.

To implement these S vectors, we reused software that already existed in the Fortran library. A set of Fortran routines supported data manipulation for some Bell Labs applications with dynamically allocated arrays of various types, among them arrays whose elements each referred to another array—the S concept of `"lists"` as vectors of vectors.

List processing languages, notably Lisp, came from a very different perspective, where the fundamental operation was navigating the list as a graph-like structure. These languages also tended to have recursion as a basic programming tool. Fortran did not originally allow recursion.

The main additional step to support specialized list objects such as the regression fit was to allow vectors with named elements (or "components"). In the regression case, the two vectors of coefficients and residuals are the natural components. In the terminology that would later come with OOP, these are now *properties* of the regression object. In the initial version of S, however, there was no particular distinction between such an object and a list whose names were specific to the particular data, such as a list of data organized by the names of cities, departments or some other hierarchical variable. This was now object-based computing but not yet object-oriented programming.

From the user's perspective, the `reg()` function has now taken responsibility for the detailed structure of the object. If the object returned has been well designed, the user will only need to learn as much of the structure as is relevant for the current application. In fact, `reg()` returned an object with 9 components, of which `"coef"` and `"resid"` were those the user would access most often.

An object-based view of the fitted model and a functional view of computing stimulate the user to examine and perhaps modify the analysis:

```
coef(fit); resid(fit); abline(fit)
predict(fit, newData)
newFit <- update(fit, newData)
```

The first two functions are typical of basic data abstraction: Extracting information functionally gives a user view that is not dependent on the detailed structure of the object. Functions can also summarize the object or display it visually as `abline()` does, apply it to new data as `predict()` does, or update the model to add new data.

Not all of these auxiliary functions were present with the original version of S. Although the original `reg()` function has long been superseded, its successors have had the same functional, object-based form.

This stage in the evolution takes us to the mid-1980s. S was licensed and used by a modestly growing community, with particular appeal to researchers in statistics, usually based in academic departments. There were also users in some non-university organizations and a few third-party resellers. Books on S itself [3] and on extension through the interface language [4] described this version. Use within Bell Labs and AT&T generally had also grown, with several small but keen development groups being particularly active.

S as distributed both internally and externally was Version 2. The first version had been implemented on hardware and for an operating system that were never likely to be used widely. Portability was a challenging prospect, solved a few years after the initial work on S by the appearance of a portable version of the UNIX system, thanks to work by our computer-science research colleagues at Bell Labs.

Portability was achieved by re-defining the term to mean implementation under UNIX, resulting in Version 2 of S. A number of changes were made, mostly internal, but the user's view and the Fortran-based perspective remained largely unchanged.

Research on software, like all scientific research, is not independent of the people involved. At this time, the two main authors were on somewhat divergent career paths, although our technical interactions remained as happy as before. For a couple of years, I left research to head a department that was charged with developing novel software to support some of the business organizations at Bell Labs and AT&T. From a combination of a different perspective and the influence of some recent ideas in computing, this resulted in early work on a different system, labelled a *Quantitative Programming Environment* or QPE [9]. The middle name of the language was important: While remaining an interactive system in the spirit of S, QPE emphasized programming in the language itself. Stimulated in part by the philosophy of functional programming (Section 1.5), functions in QPE were objects that could be defined in the language.

Meanwhile, a number of changes to adapt S more effectively to the UNIX system and its C-based software were in development. Would this result in two distinct future paths?

Events on a much larger scale then intervened. These were years of fundamental change at AT&T. Following an anti-trust suit by the federal government, AT&T agreed to split, divesting itself of its local telephone subsidiaries. In the process, Bell Labs also split off a research organization to be jointly owned by the new regional telephone companies. In the resulting changes, I came back to statistics research at (now) AT&T Bell Labs.

Ideas were merged and a project began that resulted in a new version of S, the version that would later be the model for R.

## 2.3   Functional, Object-Based S

Once a decision had been made to create the "New S", the implementation proceeded fairly rapidly, in effect merging work on QPE and on modernizing S. By 1988, the new S was being distributed and described in *The New S Language* [5], which became known as "the blue book" from the color of its cover. This was Version 3 of S, or S3 for short. It was *not* back-compatible with previous versions. This was the only time (so far) that a version of either S or R has been fundamentally incompatible in the sense that the majority of the programming done for the previous version would not work with the new version.

In retrospect, we probably got away with the wrenching change in part because the S community, although it was growing, was still much smaller than it would become, and minuscule compared to the current R community. We were research statisticians at Bell Labs and therefore not so accessible to irate users as at a commercial software organization. In contrast to R, S was not open-source. The conservatives who wanted to stay with Version 2 would not have been able to continue with back-compatible upgrades of it. It's questionable whether a similarly fundamental change to R would be feasible now, without much more attention to back-compatibility.

Our arguments for ordinary users to switch to new S were mainly:

1. New S was just *better* for writing code, in effect because of its object-based, functional form. We didn't yet express the $\boxed{\text{OBJECT}}$ and $\boxed{\text{FUNCTION}}$ principles directly, but they were implied.

2. The implementation was equally modernized, based on C rather than Fortran except for numerical algorithms better suited to Fortran. It was also better integrated with UNIX, making it faster and adding many UNIX-style features.

3. There were many specific new capabilities on a number of levels. In addition, we had plans for more to come.

However, the essential change was not in ordinary interactive use but in programming. The new S had adopted QPE's main feature of programing in the language with functions as objects. We anticipated that ease of programming would prove to be a decisive advantage and this turned out to be the case.

Version 3 of S does not need to be described in detail; essentially, it was "pretty much like R", particularly for the purpose of interactive data analysis as opposed to programming. We'll consider some relevant differences when R itself arrives in Section 2.4.

An additional fundamental change was to drop the interface language. In the blue book, we stated that it was "no longer needed". In retrospect, that was somewhat of an exaggeration. The new interface functions, `.C()` and `.Fortran()`, called subroutines in the corresponding languages with arguments interpreted as vectors of basic data types. Fine for algorithms in those languages operating on such data and producing corresponding data as results.

Missing was the ability to do S-like programming in the interface. A response would be the `.Call()` function in a later version of S, which allowed manipulation of S objects in C. In this and other respects, S still had some evolving ahead, although less dramatic than the change to Version 3.

In writing the blue book, we made a decision to downplay the statistical content, both for reasons of space and to encourage non-statisticians who had an interest in computing with data to experiment with S. (The book still ran to 700 pages because we retained the detailed documentation as a 300-page appendix.)

Omitting statistical topics from the book was acceptable for another reason as well. A number of us had a vague, ambitious notion that we should rethink some major areas of statistical methods from the perspective of functional, object-based computing. The most attractive topic after initial discussions was the specification, fitting and examination of models for data. A number of Bell Labs colleagues and their collaborators were doing research into new types of models and the computational approaches to classic models.

The functional, object-based computations in S provided a new perspective on statistical model computations, with a potentially unifying view over a variety of models. Realizing these ideas in practice led to a challenging but very rewarding joint effort among 10 authors and eventually to software added to S and a book, *Statistical Models in S* [12], inevitably called the "white book" from its cover. This book and the blue book are relevant in that they were the public description of S used in the implementation of R. We'll look at the result in Section 2.4.

The version of S in the white book was an extension of the previous one, backward compatible but with several extensions. (We had an aversion to version numbers with S, relying on dates: The white book version date was 1991.)

The critical extension for the whole statistical models project was the introduction of an informal object-oriented programming. This was adopted in R. Section 2.6 discusses this step in the evolution. Statistical models remain as convincing an example as any of the usefulness of functional OOP, and as a result they will reappear several times in our discussion of that topic. Other changes in the 1991 version of S were not documented in the book, although some of them were mentioned in the Preface.

None of the other, less specific ideas for presenting areas of statistical methodology from the new computational perspective were pursued. The thousands of modern R packages do treat many areas of statistics, from a variety of viewpoints.

The remaining evolution of S pursued several directions, described in the 1998

*Programming with Data* book [10]. One of the main additions again concerned functional OOP; this time, an extended formal version. The version of S described there has come to be known as S4, but was again essentially back-compatible as far as the language itself was concerned.

Two circumstances complicated the reception of these new ideas. First, for several years S had been distributed under an exclusive commercial license with the (then) MathSoft company, as S-Plus. The distributed version had some differences from the internal one described in the book, and would not necessarily adopt all changes.

The second, more epochal outside event was the publication of an article entitled "R: A language for data analysis and graphics" [23]. By 1998, R had arrived and was on its way to its first official version. The evolution of S had become entangled with that of R.

## 2.4   R Arrives and Evolves

Public awareness of R largely began with the publication of the 1996 article [23]. Ross Ihaka and Robert Gentleman had begun work on the software in 1993 at the University of Auckland. The early history is described informally in an article on the R website [22]. The early form was influenced by Lisp, as can still be seen in the source code. A fateful decision however was to follow the syntax of S. As Ross Ihaka writes in the history, that effectively ensured that much of the content, not just the syntax, would replicate S.

From a user's perspective, seeing something that looked like S naturally led to the expectation that entering an expression modeled on an example in the S documentation would produce a similar result. So to speak, if it looked like a duck it was expected to behave like a duck.

Two other early decisions by the original authors were likely crucial. First, the new software would be open-source, using an already well-tested license. Second and most important, ownership and management of R would be shared with a growing international group of volunteers, soon to be labeled "R core".

By 1997, the joint project was well underway. In [22], Ross Ihaka lists 11 members for R core, not a large number for such a project, particularly given that all were volunteers, "contributing when and as they can".

Replicating S was indeed no small commitment. Not only had work on S been continuing for two decades at this point but, as I emphasized earlier in the chapter, S itself started with a substantial library of supporting software. At least the major capabilities of this library had to be provided.

Another question was what, precisely, "replicating S" meant. Unlike C, for example, no standard version or definition existed. The blue book and white book had to stand in for a definition, but those were written as user guides rather than formal documents, although the blue book in particular had tried to be fairly complete. I was not in R core at this time, but can imagine not a few occasions of frustration with the vagueness of entries in the books. When contacted (which was not often), I did try to clarify some of what we meant but as in all open-source replications, the new project needed to maintain distance from the proprietary model.

To complicate matters further, the new version of S was developing during the same time frame. What to do about features described there?

From the perspective of the present book, these events are mainly relevant to the extent they influenced the eventual content and organization of R. Version 1.0.0 of R implemented a large fraction of the functions in the blue book. From the white book on statistical models, R took the computational tools (data frames, formulas and S3 classes) and five of the seven types of models. Alternative software developed then or later for tree-based models and nonlinear models. From S4, R took some ideas, modified some and omitted others (see Section 2.7).

Back-compatibility with S was especially relevant for the behavior of individual functions, and is reflected by the citations to the blue book and the white book. Some of these constraints are probably unfortunate, but were justified in gaining user confidence in the early period.

In relation to extending R, some carryovers in the general computational model have implications. For example, the question of how to incorporate existing functionality into new extensions puts the R concept of package namespaces somewhat in conflict with the S-based notion of a search list. We will discuss these points as they come up, with some suggestions. On balance, the problems can largely be avoided and overall the pluses of the evolution exceed the minuses by a considerable margin.

The official R version 1.0.0 was dated February 29, 2000[3]—the CD shown in Figure 2.3 (one of my treasured possessions) is evidence of the event.

For understanding R, it needs to be repeated that creating an "open-source S" was not a routine exercise or simply an implementation according to specification. The blue book and white book established the user's view, if sometimes vaguely. Supporting it all, the original subroutine library used by S incorporated many of the available state-of-the-art algorithms (for numerical linear algebra, for sorting and searching and for random number generation, among others). These were still available, or replaced by improved equivalents.

---

[3]A cleverly chosen date. It was the first use of the third-order correction to the Gregorian calendar: a day is added to February every 4 years; the second correction cancels that every 100 years; the third restores it every 400 years.

Figure 2.3: A compact disc containing the first version of R. Serial number 1, given to the author by the members of R core.

The result for R was somewhat of a "sandwich", with familiar top and bottom layers but a filling that contained new ideas and solutions. Among these were some key ingredients in the result, which will feature in our discussion here. For example, environments in R are an essential tool in implementing both the OBJECT and FUNCTION principles. S had no corresponding basic type.

Once launched and officially available, R increased rapidly in user base, areas of application and contributed software. The contributions and their value to the user community were facilitated by the R package mechanism. This formalized and facilitated adding new software, both R functions and other tools, notably via interfaces. The CRAN repository was the most important central maintainer of such packages, as with similar repositories for other languages such as Perl and Python. The package mechanism has become the key step in extending R, and will be a focus of much of our discussion in Part II.

The total picture—packages, user interfaces, the user community worldwide and all the contributions it has produced—make for a resource unprecedented in statistical computing. The varied benefits from these contributions justify our attention to extending R as an important goal.

# 2.5 Evolution of Object-Oriented Programming

The techniques of object-oriented programming can be very helpful in extending R. It is arguably the most effective way to manage change and complexity in the type of applications we need to pursue. While OOP fits well into the overall structure of R, as I will argue, it comes from a different stream of development than we have considered so far. To put it into our evolutionary perspective, we need to step back temporarily from considering S and R directly.

Functional objects in early versions of S had only informally defined structure. In the regression example in Section 2.2, page 27, whatever the function `reg()` returned, that was the regression object.

Contemporary with the first versions of S, but generally far away from data analysis, ideas introduced in a variety of contexts gradually converged on the notion of *classes* of objects. Languages were introduced that allowed programmers to express their computations in terms of such classes. The Simula language was an early pioneer; as its name suggests, a main goal was to program computer-based simulations.

In Simula and other languages following, such as C++ and Java, objects that belong to, or have, a designated class behave similarly, using the concepts outlined in Section 1.6. The objects will have certain known *properties* that can be queried and, usually, changed. Computations on objects from the class will be able to invoke known *methods* on the object.

For the class to be an effective way of organizing computations, objects from the same class should have comparable properties and methods should work on all objects from the class. A complete definition of the class in object-oriented programming includes the definition of properties and methods.

Simula also introduced the concept of *inheritance* or subclass: A class could be defined as a subclass of an existing class, inheriting the properties and methods of that class but potentially adding or redefining some of them. This concept has proven to be very valuable: A refinement of a class can add or modify just those properties and/or methods that distinguish the subclass, while continuing to use all the methods that remain appropriate.

All the objects in any subclass have the properties of the original superclass, but each of the subclasses may add properties not shared with the others. Methods defined for the superclass can be used with objects from the subclass, since those objects will formally have all the properties referred to in the method. Usually, however, one of the reasons for defining the subclass is to redefine some of the computations. In this case, the subclass may override the relevant methods, as well as introducing new methods of its own.

Although Simula is apparently still around, it is not one of the languages we will discuss explicitly. Except when talking about a specific language, we will present examples as they would be written in R. Simula and other languages use the character "." as an infix operator for properties and methods. Since that character is used in names in R, the operator `` `$` `` is used instead.

As an example involving simulation, suppose we wanted to model a population of organisms that evolved through time. Let's define an extremely simple version of a class. Each object from the class represents one simulation of such a population. Objects from this class have three properties: the object has a vector `sizes`, recording the number of organisms in the population at each discrete time during the simulation. Each organism has some probability, `birthRate` of splitting into two and also a probability `deathRate` of dying at each time in the simulation. As a result, the population may grow or shrink.

Suppose `"SimplePop"` is the class (an implementation in R is shown on page 143). Let's create an object corresponding to one realization of the simulated population, with chosen initial size, birth and death rate:

```
p <- SimplePop(sizes = 100, birthRate = .08, deathRate = .1)
```

And now we take the population through as many evolutionary steps as we want, each of which is obtained by invoking a method:

```
p$evolve()
```

The software keeps track of the population size after each evolutionary step; `p$sizes` contains the vector of population sizes after each step taken since the object was created. The method call `p$size()` returns the current size. Evolution continues for 50 steps, unless the population dies out along the way:

```
> for(i in 1:50)
+    if(p$size() > 0)
+       p$evolve()
> plot(p$sizes)
```

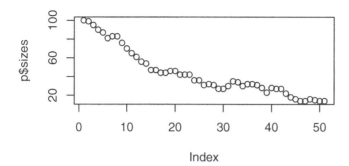

The object `p` from class `"SimplePop"` and an object `fit` from a modern version of linear regression both represent models but they differ fundamentally in how we think about and use the object. The difference is at the heart of the two paradigms for OOP that will occupy us throughout Part III of the book.

An object from the class `"lm"` of linear regression models is returned as the result of a call to the function `lm()` (the generating function of the class, in OOP terminology). Its functional nature is essential. To be trusted, it must be a valid result of a well-defined functional calculation.

A new object could be created by changing some values in the `coef` or `resid` components of `fit`, but it would have to be carefully constructed to be a valid member of the class. The generic function `update()` is designed to do just that; namely, to create a valid linear model representing certain changes from an existing model. It has a method for class `"lm"`, but that method does some careful, functional calculations and returns a new object from the class as its value. Arbitrarily tweaking properties of the object does *not* leave the object the same or even valid.

The population model, on the other hand, is an object that is created, studied and changed—it evolves. In the usual terminology of OOP, such objects have a *mutable state*, that is, internal properties that can be examined and potentially modified by calling a method, here `$evolve()`. The internal state of the object changes, but it is still the same object, in contrast to the functional case.

The three major early languages—Simula, Smalltalk and C++—had distinct focuses. Simula dealt with simulation, more from the perspective of engineering and physics than that of statistics. Smalltalk was the basis for a pioneering programming environment that introduced now standard modes of interaction, such as multi-window terminal displays and mouse-oriented, menu-based user interactions. C++ began at Bell Labs as "a better C"; as such it was used for basic computational tasks, leading to major software in many fields including combinatorial (e.g., graph layout) and numerical (e.g., optimization) libraries. Simula and Smalltalk, while of great historical importance, are not currently relevant. In contrast, C++ has continued to evolve and be applied in many areas of importance to computing with data.

Subsequently, additional languages adopted versions of the paradigm. Some, such as Java, did so from inception. OOP is at the heart of programming with Java. Other languages have added OOP techniques, typically to languages that initially had a more casual and less demanding programming style. Python and Perl are examples. The implementation is not always complete. Class definitions in Python, for example, do not include the specification of properties.

Simula and C++ added OOP concepts to an existing procedural language (Algol60 and C, respectively). The additions are principally class definitions, including methods associated with the class, plus an operator in the language to invoke the methods.

A number of languages with encapsulated OOP are among those valuable for interfaces from R. Part IV of the book will discuss these generally in Chapter 12 and specifically for Python and C++ in Chapters 14 and 16.

As an example of the style of encapsulated OOP, here is a C++ implementation of vector convolution, a computation we will use to illustrate interfacing from R and also to discuss some efficiency questions (page 91, for example).

```
NumericVector convolveCpp(NumericVector a, NumericVector b) {
{
    int na = a.size(), nb = b.size();
    int nab = na + nb - 1;
    NumericVector xab(nab);

    for(int i = 0; i < nab; i++)
      for(int j = 0; j < nb; j++)
        xab[i + j] += xa[i] * xb[j];
    return xab;
}
```

Someone familiar with C but not with C++ could read the code without difficulty, except for expressions such as `a.size()`, where the operator `"."` *invokes* the `size()` method *on* the object a. Also, what appears to be a C declaration for type `NumericVector` takes parameters in round brackets, function-style, as opposed to square brackets for C extents. In fact, `NumericVector` is a class defined in C++, and the apparent declaration also invokes a constructor at run time, possibly with several optional argument types. The definitions of the `size()` method and of the constructors are encapsulated in the class definition, or in the definition of a superclass. At the same time, the code contains non-OOP type declarations (`int`) and procedural expressions (`nab = ...`). An idiosyncrasy of C++ would not be obvious to the C reader: Operators in the language, such as `` `[]` ``, can be overridden by special method definitions. The key computation inside the double loop is in fact using such hidden methods.

The C++ class `"NumericVector"` is not particularly elaborate, but its defined methods and overloaded operators allow simpler implementations of suitable computations. The name of this particular class is not accidental: it provides a convenient and efficient interface for computations with R numeric vectors (Section 16.3, page 345).

# 2.6 Functional OOP in S and R

The version of S described in the "blue book" and replicated in R emphasized programming directly in the language, and in effect adopted the OBJECT and FUNCTION principles.

This was still *object-based* computation, as distinguished from *object-oriented programming* with the concepts of classes, methods and inheritance. The first organization of S objects into classes came as part of the software described in *Statistical Models in S* [12].

In the blue book, a function `lsfit()` took the place of `reg()` with some extensions but essentially the same organization. The function `lm()` in the white book also returned a similar object, but intended to be part of a more general computational framework. In addition to linear regression, the software included analysis of variance models, generalized linear and generalized additive models, tree-based models, smoothing techniques and nonlinear models.

From the user's perspective, three features were emphasized.

- Model-fitting was object-based, in that both the arguments defining the model and the result returned from fitting it were objects.

- Flexible display and analysis of the fitted models was emphasized, with special emphasis on graphical display.

- The functions that generated the fitted model objects and also those that did computations on the resulting objects were organized uniformly so that the user's programming carried over from one type of model to another.

A linear regression, for example, with the classic `iris` data:

```
irisFit <- lm(Sepal.Width ~ . - Sepal.Length, iris)
```

The two arguments are a formula describing the model and a data frame that supplies the variables.

Other types of model would be generated by functions such as `glm()` for generalized linear models, `gam()` for generalized additive models and so on. The formula and data arguments would be similar, with optional other arguments to specify additional information that the particular type of model might require.

The object representing the fit then became an argument to functions extracting information from the model, producing printed summaries or graphical displays, or computing predicted values or updates of the model corresponding to new data. For the user, it was desirable to learn just one way to produce each of these additional computations, regardless of which kind of model was involved:

```
coef(irisfit); plot(irisfit);
predict(irisfit, newFlowers)
```

The same `plot()` or `predict()` function should produce suitable graphical displays or predicted values given objects returned by `glm()` or `gam()`. The computations for the plot or the prediction, on the other hand, would likely need to be specialized to the type of model. A computational challenge was to implement these variations while keeping the organization of the software clear, particularly when different subsets of the ten authors would be implementing the different model types. Adding code to `plot()` or `predict()` for each type of model would quickly make the functions unmanageable.

We realized that object-oriented programming was the way to proceed. Computations such as the `plot()` and `predict()` functions were "embarrassingly OOP": clearly the user wanted the functions to use the appropriate method based on the class of the model object. The structure of the fitted objects could help: `coef()` might find the information by some standard structure. The `plot()` and `predict()` functions needed specialized computations for each class. The linear regression function, `lm()`, returned an object of class `"lm"`; `plot()` and `predict()` should call a corresponding function as a method.

Although we had now taken on the essential OOP concepts, those concepts had undergone a large transformation in the process, relative to those in languages like Simula. The objects were created by functions such as `lm()`. The user called other functions, with arguments that needed to be as uniform as possible for all fitted models.

What we had created was a form of OOP suited to functional computing, *functional OOP* in other words. We did not entirely grasp the distinction ourselves, and many years of comment has demonstrated that users of standard OOP usually did not grasp the distinction either.

Our first version of OOP in S was lean and simple. Classes had no definition.

Method selection was determined by matching a function name to a `"class"` attribute, in principle arbitrary for each object; so, `plot.lm()` would be the plot method for class `"lm"`. Methods were selected based only on the first argument to the function, aside from a few special hacks. This version of functional OOP, usually called S3 classes and methods, is still in active use. Initially described in the Appendix to [12], it is discussed here in Section 10.8.

Further evolution in S and now in R has produced a more formal approach to OOP, usually labeled as S4. Classes have explicit definition, with properties of specified class. Methods are formally specified and are kept in a table in the generic function during evaluation. Methods have a signature specifying a class for one or more formal arguments.

This is the version of functional OOP we will consider in this book, with a programming introduction in Chapter 9 and an in-depth consideration in Chapter 10.

## Example: Relational databases

To give the flavor of functional OOP, here is an example from one of the early applications, interfaces to relational database management software.

**The goal:** To provide software giving convenient access in R (originally, in S) to data in any of a number of relational database management systems.

Relational DBMS are a collection of widely used systems, both commercial and open-source, for managing data using the *relational model*. Data are represented conceptually as tables, with named variables defining "columns" in the table. The natural analogue in R is the data frame. Data is generally accessed via commands in the SQL format, dating far back in computing.

Implementations of the relational model include commercial systems from Oracle, IBM, Microsoft and others. Open-source systems include MySQL, PostgreSQL and SQLite. Similar DBMS that are not strictly SQL include MonetDB.

Although SQL and the systems implementing it are used for all phases of creating, updating and querying databases, the natural focus for an interface from R is on querying with the intention of performing some analysis on selected subsets of the data.

The goal is an interface including R functions that communicate queries and potentially other commands in a standard, natural form. The functions will accept SQL statements as character strings and will communicate with the DBMS software through R objects that convey the relevant details of this particular database; the user's view will be as free of these details as possible.

Once stated in this form, the database project is, like statistical models, "embarrassingly OOP". The functions called by the user will be generic, in that their purpose and calling sequence will not depend on the particular system used. To specialize the functions, a class of objects will correspond to a particular database system. Methods will be defined for the generic functions and the class as needed to make things work.

The database interface in this form was one of the first projects to use S4 classes and methods. David James created the first version in S, and later ported it to R as the DBI package, [30] and [11, Chapter 12], still actively used and maintained now by the R special interest group on databases.

The user will create database connections using a particular database system, through software in an R package which is written for that system and which provides classes and methods extending those in DBI.

For example, SQLite is a simple, widely used system for storing relational data in files. The RSQLite package [39] provides an interface. To communicate with an object in an SQLite database, the user opens a connection object for the particular database. The class of the connection object identifies this as an SQLite database.

Queries and other database computations are generated by calling the generic functions in DBI, with a database connection object as argument.

The functional organization and the use of OOP allow multiple levels of involvement for the user. At the simplest level, a simple call reads a whole table from the database and returns a data frame. The SQL details are transparent to the user.

If more control is needed, as for example with a large table to be processed in blocks, the user needs to understand something about queries in SQL. Other classes in R represent the intermediate objects needed for the query. The user needs some expertise in SQL, but details of the individual DBMS's implementation remain largely transparent.

The central idea here goes back a long way and is still powerful: abstract out the common features of computing tasks; implement those as functions; make the objects that specialize the tasks arguments to those functions. Without the explicit OOP connection, it was the idea behind the original graphics subroutines [2] that gave rise to S and R base graphics. In this case, plots and other graphical displays provided the functions and particular graphic devices would have been the classes.

The same general approach motivates the XR structure for interfaces in Chapter 13. The functions communicate with other languages and create proxies for objects in those. The classes define evaluators for individual languages. In this case, an encapsulated version of OOP is more suitable, with individual evaluator objects providing methods for the interface.

## 2.7 **S4** and **R**

S4 methods and classes were presented in the 1998 *Programming with Data* book [10] and included in versions of S from then on. A number of other new features besides OOP software were presented in the book.

The 1998 date is significant: R was already under development at the time S4 was introduced, but had not yet reached its official 1.0.0 version. The question of what to incorporate from the 1998 revision was not simple.

Here are the main S4 novelties, along with their eventual fate with respect to R:

- Classes and methods did not replace the existing informal versions in either S or R. Starting in late 2000, S4 methods were added to R in the methods package, and have gradually become more central to development since then, but still without displacing earlier versions.

- S4 introduced `"connections"` as a class of objects to abstract input and output to files and other forms, including character vectors, with subclasses for the various forms (`"file"`, `"textConnection"` and others).

  R took up connections, with a number of later extensions and modifications. These were made S3 classes, rather than formal classes as they were in S4.

- S4 introduced the notion of `"raw"` data for vectors of bytes. It also added a class `"string"` that extended character vectors with slots used for table lookup and other dictionary-style computations with strings.

  R took up `"raw"` as an additional basic vector type, where it fits in quite smoothly. R has the advantage of a clear distinction between type (the internal representation of the object) and class (an attribute), providing it with a richer if slightly confusing range of computations, as will reappear frequently in this book.

  The `"string"` class was not picked up, and largely not needed. R has a different approach to character vectors than S (or than C programmers would expect), in that type `"character"` is not a vector of `char[]`, but of a special, internal R type, `CHARSXP`. This is used for similar purposes to `"STRING"` in some respects, for example in a global hash table.

- S4 introduced the *database* concept as a general mechanism for storing S objects by name.

  The `"environment"` type in R is a cleaner version of this.

- The *chapter* in S4 was the programming unit for a collection of S objects, and is subsumed by the package in modern R.

- S4 had a new system for online documentation, created by Duncan Temple Lang [10, Chapter 9]. There were three main features: (i) function source was self-documenting in the sense that comment lines preceding the `function` keyword were incorporated as documentation for the function object created; (ii) explicit documentation, if prepared, used the SGML markup language; and (iii) documentation was kept in S objects, in a documentation database in the chapter.

  This was an interesting approach, which might have developed well. However, the large and productive effort to bridge R with report generation and web-based information display has made a change in this direction highly unlikely.

Given that R has become by far the dominant descendant of S, the discussion in the rest of the book will always be of the R version of all the relevant features.

# Chapter 3

# R in Operation

Extending a software system benefits from understanding how it operates. The extension will then be more likely to be compatible with other parts of the system, to make use of relevant tools and functionality and to fit in with users' paradigm for the system.

R is most frequently used through an interactive application such as RStudio. It may also be part of a general interactive development environment, such as ESS running in an emacs application. Its original design was as a process initiated as a shell command and all the other user interfaces will eventually initiate such a process. The R process reads expressions from its input, parses these and evaluates each complete expression once parsed. This chapter examines aspects of the R process, with emphasis on how extensions to R should work with the process.

The computations are best understood in terms of the three fundamental principles. We begin with overviews of how the three principles translate into the key components of the R process: the objects, the functional computations and the basic interfaces. The R evaluator (Section 3.4) uses the components to carry out users' computational requests. The programming techniques for extending R discussed in Part II of the book build on these components.

## 3.1    Objects and References

Any programming language or command software that intends to work with objects must provide users with a way to refer to those objects. As we saw in Section 2.1, page 20, early assembly languages provided references to blocks of storage via character string names. Languages such as Fortran extended such names to refer to vectors or arrays, blocks with implied properties for valid data and a mechanism for indexing.

The references could be used by the programmer directly to access or to assign elements within the data. They could also be supplied as arguments in function calls, where they explicitly acted as a reference to the whole pseudo-object.

References in early languages were part of data declarations; they both set aside space to hold the data and declared the type and structure for the contents. In Fortran, for example:

```
real x(100, 5)
```

creates a data structure for holding 500 floating-point numbers and gives the programmer the mechanism for using the name `"x"` to refer to that data as a two-way array.

Dynamic languages such as R create similar data structures in response to computations. Given the basic principles of R, that means creating an object as the value of a function call. If this object needs an explicit reference, that is usually done by the assignment operator (which is itself a function):

```
x <- matrix(rnorm(500), 100, 5)
```

Evaluating the assignment does one thing only: It makes the name on the left a reference to the object on the right.

In any programming language or similar software, a reference such as a name exists only within a particular context, or "scope". In R there is only one kind of context, an *environment*. A simple assignment like the above is always evaluated within an environment—we'll say within the "current" environment. There are also some cases of implicit assignments, for example of the arguments in a function call, but they behave essentially the same way.

The key concept needed to understand objects and references in R is then:

> A REFERENCE TO AN OBJECT IS THE COMBINATION OF A NAME
> AND AN ENVIRONMENT.

(We could justifiably make this a fourth principle, but three is enough.)

We also have not said what either a name or an environment actually is. A name in R is just a non-empty character string. An environment is what's needed to make the fundamental concept work: a mechanism for maintaining a table of names and the corresponding objects.

There will be much to be said about computing with both names and environments in Part II. Environments are an important tool in extending R, particularly for making objects available in packages (Chapter 7). But the principle above is the key.

An object in R, then, is created when something happens; it can be referred to subsequently by assigning a name for it in an environment. When a name is evaluated, the corresponding reference is resolved.

R has a limited mechanism for fully qualified references, if the object is exported from a package. For example, `base::pi` is a reference to the object `pi` in the `base` package. Otherwise, an unqualified name is resolved by a search process in the evaluator.

This is done by first looking in the current environment, then in the parent environment (sometimes called in R the enclosing environment), then in its parent and so on, until encountering the empty environment.

While the lookup always follows the same rules, there are two distinct situations: a name used interactively, in effect calling the evaluator explicitly; and a name used in a function, part of a package that has been loaded into the R process. The two situations are quite different.

When computing interactively, the user's expressions are evaluated in the "global environment" provided by the R process (and accessible by the reserved name `.GlobalEnv`). The environments searched include the global environment and the exported objects of the currently attached packages.

For programming in a package, computations will take place in a function, assigned in the namespace of that package. The package implementation will control what objects from other packages are in the enclosing environment of the namespace.

The function `envNames()` in our **XRtools** package returns a vector of character string "names" corresponding to an environment and its chain of parent environments. The different results from the interactive session and from a function call will clarify what happens.

Interactively, here is the result, first from a new R session and then after a package has been attached:

```
> XRtools::envNames(.GlobalEnv)
[1] "<environment:R_GlobalEnv>" "package:stats"
[3] "package:graphics"          "package:grDevices"
[5] "package:utils"             "package:datasets"
[7] "package:methods"           "Autoloads"
[9] "<environment: base>"
```

The chain of environments corresponds to the "search list" returned by the base function `search()`, and comes from the notion of attaching libraries that has always been the model for interactive computing in R and in S. In effect, the chain of environments is R's way of implementing the search list mechanism.

Attaching a package inserts the environment of that package's exported objects into this chain, by default as the new parent environment of .GlobalEnv.
When the codetools package is attached in the usual way, the environment named
"package:codetools" is inserted:

```
> library(codetools)
> XRtools::envNames(.GlobalEnv)
  [1] "<environment:R_GlobalEnv>" "package:codetools"
  [3] "package:stats"             "package:graphics"
  [5] "package:grDevices"         "package:utils"
  [7] "package:datasets"          "package:methods"
  [9] "Autoloads"                 "<environment: base>"
```

The search list is convenient for interactive computing. Once a package is
attached, users can use the names of all publicly available objects. The name
"lm" matches an object in the stats package, "iris" matches one in datasets. For
programming in a package, however, one wants a safer and more explicit rule.
Attaching a package could override a reference with the same name in another
package later in the search list.

A package should explicitly import objects from other packages as needed
through import() or, more explicitly, importFrom() directives in the "NAMESPACE"
file. The R evaluator will ensure that references in the package's functions will be
found when the package is loaded, regardless of the current search list.

A call to a function from the package will search for names in the frame of the
call and then in the package's namespace (because this is the environment of the
function). The parent of that environment is composed of the objects imported
into the package, and the parent of the imports environment is the base package
(therefore importing from the base package is meaningless). We'll examine details
of how this works in Section 7.3.

A little function in the XRtools package, myEnvNames(), will illustrate. This
just returns the chain of environments from its call:

```
myEnvNames <- function(...)
    return(XRtools::envNames())
```

The result would be the same for any function in package XRtools:

```
> library(XRtools)
> envs <- myEnvNames()
> length(envs)
[1] 14
> envs[1:4]
```

```
[1] "<environment: frame 1: myEnvNames()>"
[2] "<environment: namespace:XRtools>"
[3] "imports:XRtools"
[4] "<environment: namespace:base>"
```

The first four environments are the relevant ones: the frame of the function call, the namespace of the package, the imports to that package and the namespace of the base package. Notice that there are 10 more environments. To the first four is added the whole search list, essentially because earlier versions of R did not enforce use of namespaces. You should not, however, assume anything about the search list in writing extensions to R. The structure described in Chapter 7 is the clear and safe one: all references to objects not in the package's own namespace and not in the base package should either be in the fully qualified form or explicitly imported by directives in the "NAMESPACE" file.

## 3.2 Function Calls

Function calls or the equivalent have been central to nearly all programming languages. The early concept was a procedure or subroutine. In its simplest form, this referred to a sequence of instructions in the machine that could be called in the sense of "jumping to" the computations which would do their thing and then "return" to the jumping off point. In addition, there needed to be a mechanism for passing arguments to the procedure and, in a function call, a mechanism for returning the result.

Through many generations of programming languages, and of programmers, this computational model was refined and elaborated. As what we'll call the *procedural* model, it remains relevant for C and other important languages today, although in a much more flexible form.

However, it is not a useful model for function calls in R. Attempts to apply it and related concepts about passing arguments tend to result in confusion and misinformation.

To understand R function calls, we start once again with "Everything that exists is an object." In particular, both the definition of the function being called and the unevaluated expression for the call are objects. A function call is evaluated in terms of these objects, via a very different model from passing simple pointers to the evaluated arguments. The differences can raise complications for some tools, such as byte-compiling, interfacing to other languages or selecting a method. But the evaluation model is essential to extending R; a number of programming techniques depend on it.

Let's start with the call itself as an object. The FUNCTION principle, that everything that happens is a function call, means what it says. All the pieces of the R language that make things happen correspond to function calls, even if in the grammar they may look like special operations such as assignments or braced sets of expressions. The `typeAndClass()` function in XRtools shows the class and the internal type for its arguments. To supply some expressions in unevaluated form, we wrap them in a call to the `quote()` function:

```
> typeAndClass(quote({1;2}),quote(if(x > 3) 1 else 0),
+                quote(x * 10))
      quote({   quote(if (.. quote(x * ..
Class "{"        "if"        "call"
Type  "language" "language"  "language"
```

The class of the objects reflects their role but they all have type `"language"` and in effect are function calls. Those that are not class `"call"`, however, don't call ordinary R functions but *primitives*, special interfaces to internal C code (see Section 6.1, page 100). Primitive functions cannot be created by users, so for the rest of this section we'll assume that the call is to an ordinary R function.

A call to an R function is evaluated by creating an environment, in which an object has been assigned corresponding to each of the arguments in the definition of the function. The value of the call is the value, in this environment, of the expression that forms the body of the function.

In the example on page 48, the function call `rnorm(500)` was used to fill the matrix with a sample from the normal distribution. The formal arguments to `rnorm()` and their defaults are shown by `args()`:

```
> args(rnorm)
function (n, mean = 0, sd = 1)
NULL
```

An environment for any call to `rnorm()` is created with three objects assigned, for `"n"`, `"mean"` and `"sd"`. In this environment, the body of the function object is evaluated:

```
> body(rnorm)
.Call(C_rnorm, n, mean, sd)
```

The body of `rnorm()` contains only a call to the `.Call()` interface to C.

In general, then, there are three steps to evaluating a function call:

1. The formal arguments and actual arguments are matched and combined to form special objects corresponding to each formal argument, referred to as *promises* in R.

2. A new environment is created for the evaluation, initially containing the formal arguments as promises. We'll refer to this as the *frame* of the call, using the S terminology.

   The parent or enclosing environment of the frame is the environment of the function definition.

3. The value of the function call is the value of the function definition's body, evaluated in the frame environment.

An important implication of this model is that the steps described above are the complete description: knowing the arguments and the function definition determines the value, and that value is all that matters about the call. This relates to the *functional programming* paradigm, introduced in Section 1.5 and discussed in detail in Section 5.1.

To see its relevance, consider a little function that replaces missing values in the data with the mean of the non-missing elements ("imputation"):

```
fillin <- function(x) {
    nas <- is.na(x)
    if(any(nas))
        x[nas] <- mean(x[!nas])
    x
}
```

We constructed the imputed data by modifying the argument, which could have been a serious violation of functional programming if it had corrupted the caller's original data. But R generally interprets the replacement operation strictly locally: only the local reference x is modified. If the same definition had been written in other languages such as Python or Julia, the body of the function would have to copy the argument first if functional programming was a goal.

The evaluation model is in principle uniform and clear, but different from the procedural model in important respects. Two aspects that are relevant for our programming purposes are promise objects and the frame as an environment.

Promise objects are constructed to support the *lazy evaluation* model for arguments. Arguments are not evaluated at step 1 in the list above, as they would have been in the traditional procedural model, but rather as they are needed in step 3.

At step 1, a promise—a very special type of R object—is constructed for each formal argument. At the C-level, the promise object has fields containing: the expression for the actual argument or for the default argument if the argument is missing; a flag indicating whether the argument has been evaluated (or *forced* in R terminology); the R object that is the value, if it has been evaluated; and a flag indicating whether `missing()` is `TRUE` for the argument. Once evaluation is forced, a request for the object returns the value stored in the promise, ensuring that the argument is only evaluated once. The promise object remains there, with its information about missingness and the unevaluated argument.

Directly assigning to an argument reference does obliterate the promise, occasionally causing surprises. For example, suppose we decided to print a message in our `fillin()` function, including the expression for the argument as returned by `substitute()`:

```
fillin <- function(x) {
    nas <- is.na(x)
    if(any(nas))
        x[nas] <- mean(x[!nas])
    message(sum(nas), " NA's in ", deparse(substitute(x)))
    x
}
```

With no NAs found, all is as expected:

```
> xFixed[,1] <- fillin(myX[,1])
0 NA's in myX[, 1]
```

But if column 2 had some missing values:

```
> xFixed[,2] <- fillin(myX[,2])
2 NA's in c(-0.48, -0.93, -0.56, -0.56, -0.94, 0.11)
```

Instead of the expression `myX[, 2]`, the call `substitute(x)` returned the evaluated argument. That's because the replacement expression reassigns `x` with the modified object, obliterating the promise. Information about the expression for the argument or whether it is missing needs to be obtained before any such assignment.

Evaluation of an argument is forced if the argument is passed to some C-level code that needs the value (such as the `.Call()` in `rnorm()`, which forces evaluation of all three arguments). A non-missing argument in a call to a generic function is also forced if the argument is needed to select a functional method.

Evaluation is forced indirectly if the argument appears in any expression that is itself forced, for example by being included in an argument to another function call. If an argument to a function is used in a call to any other function, you should assume it may be forced. Some functions treat some arguments specially to avoid standard evaluation (the first argument to `substitute()` for example). Such special cases are a nuisance for techniques such as byte compilation, but the mechanism is fundamental to the R evaluation model and a number of useful applications rely on it.

In addition to the formal arguments, the set of objects in the evaluation frame will contain the objects locally assigned during evaluation of the function's body. Computations in the body of the function can refer to any of these objects. They can also refer to objects in the parent environment of the frame, or in fact to any of the environments up the chain of enclosing or parent environments. Good package design uses only local and explicitly imported objects and the base package.

A useful technique in some programming with R functions is to define "helper" functions inline in the body of a function. These functions have the frame as their environment; as a result, they can refer to arguments or any other objects accessible from the frame. Such helper functions are useful when passed as arguments, for example, since calls to them can be evaluated in any other function.

Clearly, setting up and carrying out a call to an R function is not trivial in terms of low-level machine operations. Precise overhead estimates are difficult to generalize and nearly always not worth worrying about. An order-of-magnitude estimate is that the overhead is roughly $O(10^3)$ computations per function call. We show some related empirical results in Section 5.5, page 91. As discussed there, if a significant reduction in computing time for a function would *really* improve your work, you might look for or program an interface for key computations in a language, such as C or C++, with a more procedural implementation.

## 3.3 Interfaces

As we keep repeating, everything that happens in R is a function call. But we have also emphasized that R, like S before it, was built on a number of algorithms implemented typically in Fortran, sometimes in C. Therefore some computations happen but *not* in R, and there must be an interface from R to these computations.

These are *internal* interfaces that link to code in the R process at the C level. In fact, there are at least six forms, depending on how you count them, three designed specifically for R plus the three interfaces inherited from S.

For extending R, the .Call() interface is by far the most relevant. The internal interfaces other than .Call() are either restricted to the base package and so not available for extending R or else specialized and/or of historical relevance. One needs to be aware of them and, perhaps, to use them in special cases. Section 5.4 discusses the different internal interfaces. Chapter 16 describes programming for subroutine interfaces in the context of extending R. The many existing uses of .Call() are important for the current operation of R as well.

The rnorm() example in the previous section uses a C interface to generate data from the normal distribution:

```
> rnorm
function (n, mean = 0, sd = 1)
.Call(C_rnorm, n, mean, sd)
<bytecode: 0x7f99435acc70>
<environment: namespace:stats>
```

A C routine, C_rnorm interprets the R objects in the arguments, carries out the computation and returns an R object containing the result. The general form is:

```
.Call(.NAME, ...)
```

The ... arguments are arbitrary R objects. The first argument to .Call() identifies a C function, by its character string name or the equivalent. On the C side, each argument has the same C data type, one that intuitively means "a pointer to an R object". The return value of the C function has the same type.

The C entry point must be accessible to the R process, which is usually achieved by including the C routine in the source code for a package (in this case the stats package). R provides a registration mechanism that returns a special object referring to the loaded entry point, with additional information such as the number of arguments required. The R object C_rnorm is a registered reference to the C-level implementation (see Section 7.3, page 121).

.Call() is nicely simple and general in form. The degree of skill and amount of programming required to complete the connection to some desired computation will determine how easily the ⌈INTERFACE⌉ principle can be applied.

The .Call() interface was introduced in S4 and adopted into R. It has been widely used, providing many examples. Nevertheless, the C-level programming remains moderately challenging and there are pitfalls. For projects to extend R, interfaces are key, and among these are many potentially useful internal interfaces to subroutines. The recommended approach to most of these is through the Rcpp interface to C++, which automates much of the interface and connects R objects in a safe but flexible way (Chapter 16).

## 3.4   The R Evaluator

With objects, function calls and a subroutine interface, we have all the essential ingredients for computing with R. Now we need to understand how those come together; fundamentally, this is the job of the *R evaluator*.

R is a classic parse-eval program. In its original user interface, the program read lines of text from its standard input, parsed those and evaluated the parsed version of each complete expression. The parsed expression is, inevitably, an object.

For evaluation, there are three fundamental kinds of objects: names, function calls and everything else. The everything else objects are data (constants), which evaluate to themselves. Names (as object references) and function calls will be evaluated as described in Sections 3.1 and 3.2. As we have seen, both object references and function calls can only be evaluated if we also are given an environment in which the evaluation takes place.

The R evaluator, then, is implemented as a C subroutine with the parsed object and the current environment as its arguments. Conveniently for examining evaluation more closely, there is an analogous R function of the same name, `eval(expr, envir)`, which takes the environment of the call as the default for the second argument.

As noted, the defining characteristic of data or constants is that a data object evaluates to itself; for example, the numeric constant `1.0`:

```
> eval(quote(1.0))
[1] 1
> identical(quote(1.0), 1.0)
[1] TRUE
```

Numbers and character strings are the only data objects that can easily be input as constants in parsed text, but the evaluator deals uniformly with all data. In addition, R chooses to have a few syntactic names parsed as constants, notably `TRUE` and `FALSE`.

Names as objects have a fixed type, `"symbol"`.

```
> x <- 3.14
> typeof(quote(x))
[1] "symbol"
> typeof(x)
[1] "double"
```

The evaluator looks up names, as discussed in Section 3.1, starting from the current environment.

Function calls also have a fixed type, `"language"`, as we saw in Section 3.2. Therefore, `eval()` is schematically just a three-way switch:

```
function(expr, envir) {
    switch(typeof(expr),
    symbol = findName(expr, envir), # an object reference
    language = callFun(expr, envir), # a function call
        expr) # anything else, a constant
}
```

The functions `findName()` and `callFun()` in our pseudo-implementation of `eval()` carry out the computations for names and function calls described in Sections 3.1 and 3.2. If you look at the actual C implementation, they are analogous to routines `findVar()` and `applyClosure()`.

When called from the top level, as when evaluating an expression entered by a user in an R session, the evaluator performs an "eval-print" operation; that is, it evaluates the expression and then optionally carries out an extra calculation that (usually) prints something to the standard output.

The top-level evaluator is slightly more than `eval()`, in that the value of the top-level expression is usually displayed. The C code is in the `"main.c"` file in the R source code, in particular in the routine `Rf_replIteration()`.

In a simplified R model of the computation, the evaluator in effect calls a function defined as follows, with `expr` corresponding to a parsed complete expression and `.Globalenv` the global environment:

```
mainEval <- function(expr) {
    value <- eval(expr, .Globalenv)
    assign(".Last.value", value, envir = .Globalenv)
    if(isVisible(value))
        printValue(value, .Globalenv)
}
```

The parsed expression is evaluated in the global environment, and assigned there.

The evaluator then decides whether the result should be displayed. Although it's not strictly true, the concept is that the result of the evaluation has a corresponding "bit". The evaluator prints the value if the bit is on. The computation corresponding to `printValue()` calls the generic function `show()` for objects from a formal class ("S4" objects in R terminology), and `print()` for other objects. (S4 methods are assumed to be only optionally available.)

Only C code in the base can control the pseudo-bit for printing directly, but the primitive function `invisible()` turns it off. There is no `visible()` function, but `c(x)` will have the desired effect.

The primitive functions corresponding to various special calculations do the generally sensible thing. For example, assignments return the assigned object invisibly. The `FALSE` case in an `if()` without an `else` clause returns `NULL` invisibly.

In the actual evaluator, the computations corresponding to `mainEval()` are part of a loop that parses and evaluates expressions from the "console" and deals with errors and warnings.

# Part II
# Programming with R

Chapter 3 examined the pieces of R corresponding to the three principles as they are each realized in the implementation of R and as they come together in the R evaluator.

This book is about extending R, so our next step is to examine *programming* in this context: creating new software that extends what R can do. Programming with R can contribute at many levels, from focused solutions for small-scale needs to large projects in which computing with data contributes to the overall goals. Chapter 4 considers how R programming best responds to the varying scale and motivation of the extension.

The following chapters are topics in R programming related to the fundamental principles. Chapter 5 considers programming with functions: functional computing in the specific context of R and two areas where R has special structure (assignments and computing on the language). Function-level internal interfaces are also discussed here, including their role in considerations of computational efficiency. Part IV will consider interfaces from R in general.

Chapter 6 similarly looks at some topics related to objects in R: an understanding of how objects are actually structured; the management of dynamic memory for objects; and reference objects, particularly environments.

In any substantial project, programming for extending R needs to be organized via one or more R packages. Chapter 7 looks at the structure of packages, at the installation and loading operations that make packages available to the R user, and at options for sharing packages.

These are "topics": a complete story of programming for extending R would be a substantial book on its own, and this one is long enough already. The topics have been chosen to emphasize ideas and techniques helpful in turning R programming into a valuable extension of the software. Sources of more information will be cited as we go along, and I urge readers to look actively on their own as well. There are many good books, online forums and other valuable sources.

# Chapter 4

# Programming in the Small, Medium and Large

One of the attractions of R has always been the ability to compute an interesting result quickly. A key motivation for the original S remains as important now: to give easy access to the best computations for understanding data.

Because the same language is used for direct interaction and for defining functions, the step from computing to programming is barely a step at all. Introductions to R should quickly encourage users to write their own functions.

In this book we are interested in extending what R can do, a more substantial goal. The R community is diverse and the projects of importance to its members range on all scales and motivating goals. As the scale of a project grows, the tools that helped at the smaller scales remain useful while some new tools become important.

In a discussion of programming with R it helps to distinguish three scales: small, medium and large.

- Programming in the small is turning an idea into software in order to try out the idea quickly while it's still warm in the mind.

- Medium-scale programming arises when the software is likely to be reused and carries out some related and reasonably well-defined computations. Therefore, its design is worth some effort.

- Programming in the large is the whole software effort required to deal with an important project, involving as wide a range of techniques as needed. The goal is to satisfy some human or organizational need, with programming a means to this end.

Where learning from data is part of the goal, R can be helpful at all scales, and at all scales valuable extensions to R can result. At each scale, some aspects of R are particularly important.

## Small

Because the same language is used for interaction and for programming, programming in the small just means reusing ordinary computations. The famous or infamous first-program example, "Hello world!" is not a programming task at all:

```
> "Hello world!"
[1] "Hello world!"
```

It's no coincidence that this example comes originally from non-interactive systems where the mechanics of programming is itself something that needs learning. In R, we can consider immediately the real concerns of programming: defining the task, communicating instructions and using the result.

Programming in the small in R equals creating functions, by assigning a function object with the desired computations in the body and, usually, turning some of the inputs to that computation into arguments (the OBJECT and FUNCTION principles again).

An important advantage of R and other interactive languages is to make programming in the small an easy task. You should exploit this by creating functions habitually, as a natural reaction to any idea that seems likely to come up more than once.

As an example, let's look at a dataset called **flowers**. In fact, it's a version of the famous iris data, but deliberately salted with a few missing values.

The iris data is a classic example for clustering or discriminant analysis. We might, for example, try the **kmeans()** procedure in the **stats** package.

```
> kmeans(flowers, 3)
Error in do_one(nmeth) : NA/NaN/Inf in foreign function call
  (arg 1)
Calls: kmeans -> do_one
```

This slightly obscure error message is saying that the computation can't deal with missing values.

```
> apply(flowers,2, function(x)sum(is.na(x)))
slength  swidth plength  pwidth
      3       1       2       0
```

The usual advice in this situation is to omit the rows with missing values. R has a convenient function, **na.omit()**, to do just that. This works, although a little experimenting shows that we should ask for several random starts to the algorithm to reproduce the "obvious" clustering into three species:

```
> fit <- kmeans(na.omit(flowers), 3, nstart = 3)
```

A disadvantage of using `na.omit()` is that we get no assignment for the observations that had missing values, even though only one variable was missing in most cases. To get an assignment of the observations with NA's, perhaps we could replace the missing values with an estimate using the non-missing data, say the mean value. A direct implementation would edit each of the columns with NA's:

```
> nas <- is.na(flowers[,1])
> flowers[nas, 1] <- mean(flowers[!nas,1])
```

But a functional approach is a better idea for several reasons. Most obviously, we really didn't intend to alter the data permanently, but only to see what happens if we give the modified data to the model computation. The functional form in which we used `na.omit()` is the natural expression of the computation in R.

Both the FUNCTION and OBJECT principles apply: it is in the spirit of R to have an object that represents the adjusted data and to have a function that maps the original object into the desired one.

The same computation in a functional form might be:

```
fillIn <- function(x) {
    for(j in 1:ncol(x)) {
        nas <- is.na(x[,j])
        x[nas,j] <- mean(x[!nas,j])
    }
    x
}
```

Now we can just substitute `fillin()` in place of `na.omit()`.

```
> fit <- kmeans(fillIn(flowers), 3, nstart = 3)
```

This also gives the expected clustering. Looking at the first two variables and labeling points from the cluster model above:

```
> plot(flowers[,1:2], pch = as.character(fit$cluster))
```

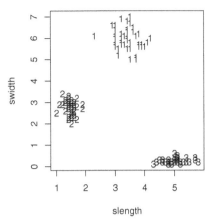

The XRexamples package has both the data and the `fillIn()` function.

Defining a function has turned the example into a small extension to R, at least locally, and improved the analysis. The function makes the general strategy clear, not obscured by the particular properties of this dataset. The function can be used again. It also suggests further exploration, for example by making the summary function an argument with `mean` as the default.

Perhaps the most important contribution of functional computing in small-scale programming, however, is to our thinking. Data analysis *is*, in a fairly deep sense, functional: we are constructing models, summaries, displays and transformations as functions of the data. In this little example, our `fillIn()` function is a simple form of *imputation* for missing values. Packages and books have dealt with this (for example, the Amelia package on CRAN [19]). We might go on to investigate these as alternatives, if missing data was a serious issue.

A functional computing approach to programming in the small helps in the short run and in leading us to a larger scale view if that turns out to be desirable.

## Medium

The medium level in R programming comes with the recognition that the new software has some lasting value. With this recognition (or sometimes, even earlier) there is usually a broadening of horizons. What other kinds of data would suit this computation? What are alternative computational methods? Are there interesting ways to display the results? This is also a good time to ask the question "What existing software in R could be helpful?", assuming that question has not been asked before.

Given the huge and diverse collections of extensions to R now available, some serious searching may answer that last question thoroughly enough to discourage pursuing our own programming. That might well be the case with the little `fillIn()` example above. A decision that existing software is all that's needed does *not* mean that the preceding programming-in-the-small was a bad idea. The steps of formulating and trying out ideas, particularly in functional form, are often essential to crystalize the questions and focus thinking.

A decision that the programming project is worth pursuing, on the other hand, is essentially to say that this is an extension to R, whether we anticipate making it public or only adding it to the tools for the local enterprise. In that sense, nearly everything in the rest of the book discusses ideas and techniques that I hope will be helpful. There are other useful techniques as well, of course, that should be mentioned.

For programming with R, the transition to the medium level nearly always benefits from creating an R *package*. This step is sufficiently important that all the discussions in the book aimed at programming for substantial extensions to R will assume that an R package is the context. Chapter 7 has some arguments for the importance of packages, if you need convincing.

Admittedly, a package has more structure and requires more preparation. Typically, some existing R function definitions and/or data are used to create the initial source package. Tools in base R, in packages such as Rcpp and in development environments such as RStudio can help.

Some additional activities, beyond simple programming, will be needed:

- *Sharing.* Making the package available, either locally or publicly, involves installing it in some form that is accessible to others. Section 7.4 discusses two of the available mechanisms.

- *Documenting.* Users need documentation. In addition to conventional R documentation, some of the many other communication mechanisms that exist today may be valuable. Chapter 7 discusses a few topics in this area, but there are many more useful tools. The entire area of web-based presentation of R software has been omitted here, but is worth exploring.

- *Testing.* The greater importance of software at this scale implies more concern for its quality. For many people involved with computing, this is assumed to be more or less equivalent to testing and in particular to having a set of tests that are complete in coverage and thorough in the aspects tested, as far as possible.

  Testing in this sense is not a topic covered in this book. In the R package process, testing is essentially checking and in particular the "R CMD check" utility (Section 1.3 of the *Writing R Extensions* manual), which performs a set of checks on an entire package and is in some respects designed from the viewpoint of validating packages when they are to be installed in a repository.

A different view of testing focuses on tests targeted at particular "units" within the package or other software, such as functions or classes in R (hence the term *unit testing* often used for this approach). R packages exist to help test specific units, notably the RUnit package [7].

Test scripts in either form are an important addition to the package. An exclusive or excessive reliance on testing, however, suggests that no one has a thorough enough understanding of the software to have confidence in it. This situation may be unavoidable for large, complex systems such as a space probe. For extending R through a package, testing should be a supplement. The software itself should be understandable and convincing. Techniques such as these will contribute to that goal:

- *Importing.* An R package should show clearly what other R software is being used. Modern R packages can be explicit about their dependencies, by specifying imports. Convenient tools allow the package to control these steps (Section 7.3, page 118). The best policy is to be explicit about all functions and objects used from other packages.

- *Functionality.* Packages need to exploit the functional concept emphasized for small-scale programming. The functional programming paradigm is particularly relevant, being substantially motivated to improve understanding and demonstrate validity. Because R supports but does not enforce functional programming, it is good to understand how to provide it (Section 5.1).

- *Object-oriented programming.* Another consequence of scaling up is often that the range of applications is extended, perhaps along with the range of strategies implemented. These in turn increase the complexity of the software. As noted in Chapter 1, object-oriented programming is perhaps our best tool for managing complexity.

  Functional OOP fits particularly well R package. Depending on the purpose, it may be natural to construct some related classes to represent interesting kinds of data, or to define some general (and therefore, generic) functions to produce the desired results, or both. Functional methods are then the natural implementation for the concrete properties or functionality desired. (Chapter 10).

There is much more to say about R packages and their relation to extending R. All of Chapter 7 will be devoted to the topic and you should be active in looking for other information as well, such as books (notably [37]) and the many blogs, mailing lists and other community resources online.

## Large

Programming in the large, in my categorization, is programming that serves the needs of a project with goals beyond the software itself. R is increasingly recognized as a valuable contributor when projects can benefit from data-based insights.

Extensions to R are found in various diverse projects: integrating analytic results into the workflow of an organization; organizing and analyzing massive data as part of a scientific experiment or clinical study; looking for insights in social media and diverse other areas where challenging use of data is needed.

The distinguishing feature of programming in the large in this sense tends to be the diversity of computational needs, certainly more than one or a few small tasks and often not reducible to a single focused medium-scale project either.

The programming should be preceded by some thoughts about overall software design (or, maybe more likely, some initial playing around will make us realize the need for such a design). The design process needs to propose some integrated set of tools that will support the larger goals. The software must be integrated into the whole data-based processes, often including data acquisition, analysis and the presentation of results to non-participants.

The principles and tools important at the small and medium scale remain important: functions and packages. If the larger goals of the project are important, the quality of the software should be a focus. As the project evolves, increased complexity may be imposed, suggesting object-oriented programming as one technique.

Challenging modern applications are often "large" in several respects: massive data sources, heavy computational requirements and multiple, diverse goals. A wide range of scientific and business applications have in common large quantities of data requiring non-trivial analysis to obtain the needed information.

The fundamental lesson about programming in the large is that it requires a correspondingly broad and flexible response. In particular, no single language or software system is likely to be ideal for all aspects. Interfacing multiple systems may be the best response. Part IV explores the design of interfaces from R.

# Chapter 5

# Functions

This chapter deals with some aspects of functions likely to arise in extending R. As argued in Chapter 4, programming in the small, for immediate satisfaction, essentially *is* programming with functions when using R. Extending R takes us to the next step in the scale, in particular to the desire to create some R software—the extension—that implements a relevant technique, idea or application. Functions remain central, but with an interest in how well the function accomplishes its purpose and often as part of an R package.

Section 5.1 relates functional programming to R, with emphasis on achieving and demonstrating functional behavior, a contribution to the reliability and safety of an extension to R.

Sections 5.2 to 5.4 analyze aspects of R computations that differ from many other languages and may need to be understood to program effectively.

Section 5.2 discusses *replacement expresions* that assign or modify objects in the local environemnt. R has a particularly flexible approach to such expressions. The uniformity of replacement expressions in R also contributes to its support of functional programming.

Section 5.3 discusses techniques to compute with *language objects*. The $\boxed{\text{OBJECT}}$ priniciple and lazy evaluation of arguments allow R to perform symbolic computations and to construct expressions to compute specialized results.

The $\boxed{\text{INTERFACE}}$ principle and the evolution of R mean that basic computations rely on R's interface to lower-level code in C or Fortran. Section 5.4 examines the various versions of this interface.

Section 5.5 examines some aspects of computational efficiency.

Although the discussions try to be self-contained, they are individual topics rather than a complete review of functions in R. Additional reading for that includes [36, Ch. 6] and [11, Ch. 3].

# 5.1   Functional Programming and R

The three main criteria of functional programming as a computing paradigm can be summarized:

1. *Any computation can be expressed as a function call.*

2. *The value of the call is uniquely defined by the values of the arguments.*

3. *The effect of a function call is simply the value returned.*

The goal is trust—to know that the function's results are reproducible and that it can be safely used anywhere.

The first criterion is essentially the $\boxed{\text{FUNCTION}}$ principle for R: whatever happens is a function call. The two remaining criteria essentially require that the computational function mimics the mathematical definition of a function: a mapping from one set (all possible argument values) to another set (all possible results) with no other dependencies or side effects.

In R this means that the function definition does not depend on any outside objects or internal variables that might change and that in turn it does not alter any such objects or values.

The second criterion is the most relevant for creating quality software in R, and is also the most challenging to put in practice. It says that we have a known and complete definition of this function, meaning in turn that we must have such a definition for every function that is called from this function. And we are asserting that none of these definitions is conditional on some external data that might change from one call to another.

The first stage in validating a function is just to know its definition and the definition of all the functions it calls.

(i) The function itself should be in an R package. The result of calling a function that is evaluated from a separate source file may be dependent on the search list in the R process when the call occurs.

(ii) Every function called from this one must either be in the same package or be an unambiguous reference to a function from another package.

    An outside reference is unambiguous either if fully qualified, as in the form `stats::lm()`, or if the function is explicitly imported by a directive in the package's `"NAMESPACE"` file, such as

    ```
    importFrom(stats, lm)
    ```

Any unqualified name remaining would need to be a reference to a function in the base package.

(iii) To defend a particular use of our function, we would also need to specify the versions of R and all the packages used.

The most important requirement is (ii). Repositories such as CRAN have increasingly pressed contributors to make external function references explicit in this way. It's a good practice and should be followed for extending R.

With this practice in place, we can assert precisely which functions may be called. The clarity and reliability of the new function now rests on those qualities for the functions called.

Functional programming requires that the value of the function be independent of any outside state of the system. Only then can one use the function with no doubt that it will produce the expected result when incorporated in other computations. Without this guarantee, trust in the function cannot be complete.

The most common example of outside state in R is probably the generation of "random" values from various distributions. Values from any pseudo-random generator are only reproducible if the initial state of the generator is known.

The kmeans() clustering function in the stats package fails functional computing by using random initial configurations. As a result, two calls with the same arguments can produce arbitrarily different values. Using the **flowers** version of the iris data (see page 62):

```
> set.seed(382)
> x <- na.omit(flowers)
> fit <-kmeans(x,3); table(fit$cluster)

 1  2  3
50 50 45
> fit <-kmeans(x,3); table(fit$cluster)

  1   2   3
 28 100  17
```

The two identical calls produced totally different clustering. Without modification, kmeans() could not be trusted in a functional computation.

Other non-local values are supported in R by the options() and getOption() functions. Users can query or set arbitrary named objects, effectively in a global table. These options are a convenience to the user, typically for setting a tuning parameter in a computation. But any function depending on them potentially returns different results for the same arguments on different calls.

The S3 generic function `aggregate()`, also in the `stats` package, has a method for time series (class `"ts"`) that splits a time series into subsets and returns a new time series summarizing the original one. The method has an argument, usually omitted, that defaults to the global option `"ts.eps"`:

```
> args(getS3method("aggregate", "ts"))
function (x, nfrequency = 1, FUN = sum, ndeltat = 1,
    ts.eps = getOption("ts.eps"), ...)
NULL
```

The value returned by the function could be changed by setting this option at any time in the R session before the call to `aggregate()`.

What to do? Functional programming is quite clear: Any variable parameter in the computation must correspond to an argument to the function and in R any default for the argument must be a known value. Note that we are not arguing against any of the computational techniques involved. Random initialization and variable convergence criteria are valuable. But to use them in functions that can be trusted, they must have a specification that can be understood and used to reproduce results. Functions using randomization should provide a mechanism for setting the seed via an argument and a fixed default setting for that argument. Tuning parameters should also be arguments with a fixed default. Users can set an option and then call the function with the value of that option as the tuning parameter. Functionality is then the responsibility of the caller, but the function is clearly defined.

Other mechanisms for specifying tuning parameters and control values exist, particularly when interfaces to subroutines are used. C, C++ and **Fortran** software may reference global quantities and provide mechanisms for setting them. There is no simple solution here; using high-quality software, preferably open-source, gives some confidence. Still, digging deeply may be required if the reliability of the final result is important.

The third criterion for functional programming, avoiding side effects, is also important, but relatively more straightforward to satisfy and to check than the second. Function call evaluation in R that uses normal R vectors and similar classes of objects and manipulates these with ordinary assignment and replacement operations will avoid side effects through the general R mechanism for these operations, as described in Section 5.2.

Explicit computations with side effects usually result from a non-local assignment (e.g., the `` `<<-` `` operator or `assign()` function). Also invalid are functions that set some global quantity; the `options()` function used to set an option, for example. Other functions called should be examined if their behavior might be suspect.

Reference objects such as environments and external references can lead to side effects. The reference objects most likely to occur are environments, type `"environment"`, and external references, type `"externalref"`. Environments are a particular danger because ordinary R replacement operations using the `` `$<-` `` operator will be non-local. External references are most likely to cause side effects through an interface to some other language. Operations in that language called from within our function could have arbitrary side effects if the external reference is a proxy for an object in that language.

Even reference objects will have no external effect if they are created locally and are not kept over multiple calls to the function. For example, one could create and use an environment for local storage and reference without external side effects, but not if the environment existed outside the function call.

# 5.2 Assignments and Replacements

This section considers all computations in R that create or modify objects within a function call by the use of the binary assignment operator, `` `<-` `` (or optionally `` `=` ``)[1]. These are collectively the *replacement expressions* in R.

The R version of these expressions is distinctive in several important respects from analogous computations in other programming languages. Most languages allow only a fixed set of expressions on the left side of an assignment—what are traditionally called "lvalues". R supports a general replacement expression in which the left side of the assignment can be any suitable functional form. This provides flexibility while supporting the locality of references needed for functional programming.

We begin by examining the general form and its implementation.

Replacement expressions lead to a useful programming mechanism in R, the *replacement functions* and possibly functional methods for them (page 76).

Understanding the actual mechanism for local replacements is helpful in analyzing likely storage requirements of computations doing multiple replacements (page 78).

---

[1]The specific choice of `` `<-` `` dates back to the first version of S. We chose it in order to emphasize that the left and right operands are entirely different: a target on the left and a computed value on the right. Later versions of S and R accept the `` `=` `` operator, but for exposition the original choice remains clearer.

## Replacement expressions

Replacement expressions are best defined recursively, in terms of the left-hand operand to the assignment operator. In a *simple assignment* the left side is a simple reference (a name). The interpretation is to assign the value to the name in the local environment (the frame of the current function call for assignments in a function body):

```
x <- diag(rep(1,3))
```

All other replacement expressions have the form:

```
f(target, ...) <- y
```

with `f` being a name. The reason for requiring `f` to be a name is that the function to be called is not, in fact, `f()` but a function whose name is constructed from `f`. As usual, the possible functions include operators, which are interpreted as the corresponding function name (examples below). The various levels of replacement are defined by form of *target*. The ... stands for any other arguments and `y` is any expression.

In a first-level replacement expression *target* is also a name:

```
f(x, ...) <- y
```

The name `x` must be a reference to an assigned object in the local environment. The expression is interpreted as a *replacement* in this reference. The replacement expression in this form is evaluated in R by calling the *replacement function* corresponding to `f`, which is the function named `` `f<-` ``. The value of the call is assigned as `x`. Specifically, the replacement is evaluated as the expression:

```
x <- `f<-`(x, ..., value = y)
```

The function `` `f<-` `` is a replacement function; any function with a name ending in `"<-"` and having appropriate formal arguments will be called to evaluate the corresponding replacement expression. The last argument must be named `"value"` because the call to the replacement function always supplies the right side operand by name. So, for example,

```
diag(x) <- vars
```

is computed using the replacement function `` `diag<-` ``() (in the **base** package):

```
x <- `diag<-`(x, value = vars)
```

The replacement function can do any computation. To make sense, of course, it has to return an object that would be suitable for the newly assigned x. Typically this will be the current contents of x modified in some appropriate way.

In a *second*-level replacement expression, *target* is not a name but is the valid left side of a first-level replacement; that is, a function specified by name and called with the object name as the first argument. For example:

```
`[`(diag(x), ii) <- eps
```

This would look more familiar in the equivalent operator form:

```
diag(x)[ii] <- eps
```

In a third-level replacement expression, the object name in the second-level call becomes again a valid first-level left side, and so on.

The evaluation of a replacement expression is likewise best defined recursively. A second level replacement is evaluated by two first-level replacements. In the first, the right-side value is replaced in a hidden object that has been initialized by the *extraction* function in the target—in our example diag(x). Then the hidden object becomes the right side for a first-level replacement on the actual target object, x here. In our example, using TMP1 to stand in for the hidden object name and writing both first-level replacements in their functional form:

```
TMP1 <- diag(x)
TMP1 <- `[<-`(TMP1, ii, value=eps)
x <- `diag<-`(x, value = TMP1)
```

The outer replacement function, `[`, appears only in the replacement form; the inner, diag(), appears in both the extraction and replacement form.

Higher-level replacement expressions are evaluated by repeated applications of the same computation on the expression. The analysis is done inside the R evaluator, but could be programmed in R as an exercise in computing on the language (techniques for which are discussed in Section 5.3). For a $k$-level expresion, there would be $2k-1$ functions called. At the outer level, the corresponding replacement function is called. At the remaining levels, both an extraction call to the function of that name and a first-level replacement in the equivalent simple assignment form will take place.

# Replacement functions and methods

From a programming perspective, replacement expressions are open-ended, in terms of both new functions and new methods for existing functions. They are an important aspect of the mixed functional and object-based paradigm in R (FUNCTION and OBJECT together). Classical, encapsulated OOP languages allow methods to modify the object on which they are invoked. Functional programming prohibits such hidden changes. Replacement expressions in R bridge the gap by changing objects explicitly, using local assignment to augment functional computations.

To be used in an arbitrary replacement expression, both the extraction function `f()` and the corresponding replacement function `` `f<-`() ``, must exist with both having the object as first argument and the latter having an additional final argument named `"value"`. If there are other arguments, they need to be formally identical between the two functions, because a call to the extraction function will be generated in the same form as the call to the replacement function.

For an existing replacement function, one can define an extraction/replacement pair of methods where that makes sense. For example, methods for `diag()` and `` `diag<-`() `` are supplied for various sparse-matrix classes in the Matrix package. The construction from the replacement expression will result in separate calls to the two generic functions, as the example above with `diag()` illustrates. Methods will be selected separately for the extract and replacement functions. The design of the methods will need to ensure that the two selections are compatible: the easiest way is to have identical sets of method signatures for both functions. That is not required: one of the two could specialize, so long as the method selected for the other generic does the correct computation. Check the details of method selection (Section 10.7) if in doubt.

Replacements generally correspond to some sub-structure inferred for the object. That does not have to be a simple attribute, slot or field. `diag()` is a good example. There is normally no specific attribute for a matrix diagonal. A computation infers the diagonal from the way elements are stored. For a sparse matrix class, this will be different from the usual sequence of elements in the `"matrix"` class.

To be used arbitrarily in replacement expressions, the function does have to correspond to a uniquely defined piece of the object (actual or implicit). The expression:

```
f(x, ...) <- f(x, ...)
```

with identical arguments on either side, should leave the contents of `x` unchanged; otherwise, nesting the function in a general replacement expression is likely to produce wrong results. Again, think of `diag()` as an example.

It is possible to have an intuitively meaningful replacement function, however, that can *only* be used as the outer replacement function with no corresponding extraction function.

A computation to modify an object in some way can conveniently be represented as a replacement, reducing the amount of typing and making the replacement nature of the computation clear. In this case, the replacement can appear as the outer replacement function, either in a first-level replacement to modify the whole object or more generally to modify part of an object. It does not make sense in this case for the new function to appear at a lower level in the replacement, because there will not be a consistent meaning to "extracting" with the same function.

For example, suppose we wanted to modify all or part of an object by adding some noise to it, or by some other computation. We could define a replacement expression `blur()` to do this:

```
blur(myData) <- noise
```

would add elements of `noise` to the elements of `myData`. The convenience of the replacement form is clearer if we want to operate on a piece of the data:

```
blur(diag(z$mat)) <- noise
```

adds `noise` to the diagonal of the field or component named `"mat"` in the object `z`. Without the replacement form, the user needs to extract and replace the piece of the object explicitly. There is a function `jitter()` in the `base` package that performs a somewhat similar calculation. To use it instead:

```
diag(z$mat) <- jitter(diag(z$mat), ...)
```

(Let's ignore the other arguments to `jitter()`). The indirect form is longer and more error-prone to type, as well as hiding the specific intent.

Rather than leaving the extraction version of such a replacement function undefined, it's a good idea to define the extraction function to stop with an informative error message, explaining for example that `blur()` is meaningful only at the outer level of a replacement expression.

It's also a bad idea to define such a replacement function with the same name as an existing function. Given the `jitter()` function, for example, creating a `` `jitter<-` `` would invite problems. A user mistakenly using the new replacement function at a lower level would, at best, get a confusing error message:

```
> jitter(xx)$a <- rnorm(n,0,.01)
Error in jitter(`*tmp*`) : 'x' must be numeric
```

At worst, data is quietly trashed. The user might have meant:

```
jitter(xx$a) <- rnorm(n, 0, .01)
```

and been confused by the operator form.

## Local replacement

The evaluator is responsible for maintaining the functional integrity of calls to functions, in the sense that the local assignments and replacements in the frame of a function call do not alter any object in some other environment.

The basic question is what happens when a local object is assigned, particularly in a replacement expression since these are more likely to be repeated in iterative computaions.

All replacements are evaluated by expansion into a sequence of computations that are semantically equivalent to simple assignments. The essential operation is evaluating a call of the form:

```
x <- `f<-`(x, ..., value = y)
```

In many, probably most, applications of interest the object to assign as x is a modified version of the previously assigned object; in particular about the same size. If the replacement function is behaving functionally, it will return a new object rather than having overwritten x.

This analysis suggests that *every* replacement computation that makes a small change in the contents of an object requires new storage of the same size as the current object. In particular, a loop executed n times replacing one or more elements of x would require n * object.size(x) storage.

So it would, if all the computations were done in R. But this is not the case—at some point the computations must be built-in. In terms of the base R computations, the distinction is between functions implemented as primitives, rather than actual R functions. This includes the basic data manipulations of subsets and elements, arithmetic and logical operators, standard mathematical functions and many other fundamental utilities. The replacement functions among the primitives operate on some special assumptions: fundamentally, if the object being replaced is a local reference only, overwrite it without creating a new version.

The key internal operation in the evaluator is duplicate(), a C routine called by the evaluator to copy an object when this object would otherwise be modified. To see duplicate() in action, we can use the built-in mechanism tracemem(). After tracemem(x), a line of printing is generated from then on whenever this object is duplicated. The printout identifies the object essentially by its address, allowing one to see chains of duplication.

A simple example will show the distinction. In a loop, we replace one element of a vector, first by the primitive `` `[[` `` operator, then by an R replacement function that does the same computation, in R.

```
> xx <- rnorm(5000)
> tracemem(xx)
[1] "<0x7f95edd1bc00>"
> for(i in 1:3) xx[[i]] <- 0.0 # a primitive replacement
> `f<-` <- function(x,i,value){x[[i]]<- value; x}
> for(i in 1:3) f(xx,i) <- 0.0 # a replacement in R
tracemem[0x7f95edd1bc00 -> 0x7f95edb09800]: f<-
tracemem[0x7f95edb09800 -> 0x7f95ed8ffa00]:
tracemem[0x7f95ed8ffa00 -> 0x7f95ed917000]: f<-
tracemem[0x7f95ed917000 -> 0x7f95ed8bf000]:
tracemem[0x7f95ed8bf000 -> 0x7f95ed82a800]: f<-
```

The primitive replacement effectively noted that **xx** was a local assignment only, in the global environment. On the assumption that the function was called in an actual replacement, the target object did not need to be copied.

Within the R function, the reference to argument **x** is shared, both the local reference **x** and the external reference **xx** correspond to the same object. When the primitive assignment is done in this case, the evaluator duplicates the object (the `tracemem()` calls from `f<-()`). The object is duplicated again for the assignment in the main frame.

As even this simple example illustrates, a precise understanding of the storage requirements is often not straightforward. For the great majority of applications, precisely what happens is also not relevant. However, for computations with very large data objects, performance can suffer because of storage requirements.

Understandably, users and developers may try to modify the details of their R programming to improve storage requirements, or may look for compilers or alternative versions of R to improve performance. Such efforts can be helpful, but one of my themes in this book is that a more effective general strategy for serious applications ("programming in the large") is the one suggested by the INTERFACE principle: Create the most effective and practical computation for the task, with whatever language or other software works. Then make that available as conveniently as possible from our context in R.

Section 5.5 will examine performance questions generally and have more to say from this perspective.

# 5.3   Computing on the Language

The OBJECT and FUNCTION principles, as outlined in Sections 1.2 and 1.3, mean that everything that happens in R results from function calls, and that both the functions and the calls to them, since they "exist", must be objects.

Section 3.2 related the evaluation of a function call to the combination of information in the call and information in the corresponding function definition. This section outlines how those objects can be used in computations.

Computed language objects can be adapted to save effort or to construct expressions difficult to form directly (the mechanism for calling replacement functions in the preceding section is an example). Computed objects can contain parts of the R code that generated them (think of formulas in statistical models).

Function objects can be similarly adapted to special needs. For example, the `method.skeleton()` function uses a generic function and the classes for a method to construct a call to define the method. The call:

```
method.skeleton("[", c("dataTable", j="character")
```

generates and writes out a function call object:

```
setMethod("[",
    signature(x = "dataTable", i = "ANY", j = "character"),
    function (x, i, j, ..., drop = TRUE)
    {
        stop("need a definition for the method here")
    }
)
```

The computation used the function object for the `` `[` `` operator to create a matching function call.

Computing on the language is also key to the development and testing phases of computation. Tools exist to examine and describe the software, for the purpose of diagnosing programming problems. Debugging tools can be written and customized in R itself, by modifying the contents of functions. Similarly, performance analysis can modify the software to record relevant information.

All these operations are open-ended, in that they exploit the general object structure of the computations, rather than being based on a few low-level hooks.

## The structure of language objects

A function call is conceptually a list, whose first element identifies the function to call, with the remaining elements if any being the expressions for the arguments in the call:

```
> expr <- quote(y - mean(y)); expr
y - mean(y)
> expr[[1]]; expr[[2]]; expr[[3]]
`-`
y
mean(y)
> expr[[3]][[2]]
y
```

Where one of the arguments is itself a function call, the list-like structure is repeated recursively. To examine or compute with the expression as a whole, we
must reach every subelement at every level.

The extraction of elements in vector style above is somewhat misleading. The
function call object is not what we normally call a list in R. A list object in R is
a vector whose elements are R objects. Function calls and function definitions are
implemented internally as a Lisp-style structure sometimes referred to as a *pairlist*,
because it's thought of as a pair of the first element and the remainder of the list.
If you were to manipulate these objects in C (not recommended), there are macros
to do Lisp-like computations.

At the R level, it's best to use existing tools, like those in the codetools package we will look at below. Where those are not enough by themselves, one can
convert explicitly between the pairlist object and a corresponding object of data
type "list". The usual R computations can be applied to these, and the result
converted back to a language object:

```
> ee <- as.list(expr)
> ee[[1]]
`-`
> ee[[1]] <- quote(`+`)
> expr2 <- as.call(ee)
> expr2
y + mean(y)
```

Notice that, as in the earlier example, the first element of a function call is typically a literal name. Had we inserted `+` instead of quote(`+`), this would have
inserted the function object, not its name. That's fine as an R call, but not likely
what the user wanted to see.

The elements of pairlist objects can have names. In a function call, the names
will be those in the call, if any. In function definitions the names are the formal
argument names. The internal implementation of names in pairlist objects is different from the "names" slot or attribute for ordinary R objects, but conversions
via as.list() will retain the names consistently:

```
> callRnorm <- quote(rnorm(n, mean = -3., sd = .1))
> names(as.list(callRnorm))
[1] ""      ""       "mean" "sd"
```

Because pairlist objects are rather peculiar beasts, it's best to stick to the elementary conversions in the **base** package:

- `as.list()` to convert from a function definition or a language object to a plain `"list"` vector with names attribute;

- `as.call()` or `as.function()` to convert such a list back to the internal form.

The **methods** package does not currently have much structure to help. Classes to extend language or function definition objects are fine, and used fairly often, but conversions to and from list vectors should still rely on the base functions.

## Iterating over language objects

The **codetools** package, supplied with R as one of the recommended packages, has a number of functions for manipulating language and function objects, as the package's name would suggest. We'll look here at a fairly detailed example of one application, iterating over language or function objects.

For computations on language or function definition objects, one often wants a different form of iteration, over all elements at all levels. To express this, it helps to introduce another terminology, that of *tree* objects. The object itself is associated with the "root" of the tree, the (linear) elements form the first-level "branches", the elements of each of these the second-level branches, and so on. All elements are "nodes" in the tree, either "leaf" nodes if they are elementary objects—constants or names in the case of a language object—or a subtree otherwise.

The term "tree-walk" is often used for an iteration that examines all elements at all levels. In R terminology, this amounts to applying a computation to this object and to all the elements of the object at all levels.

The **codetools** package has a basic set of functions to perform a tree walk on a language object. The tree walk itself is performed by the function

```
walkCode(e, w)
```

where `e` is a language object and `w` is what's referred to as "code walker". This is expected to be a list with some specially named elements. The element `w$call` is a function with the same formal argument structure as `walkCode()`; that is, a language object and a code walker. Element `w$leaf` is another function, also with the formal arguments `(e, w)`.

The `walkCode()` function proceeds essentially by examining the object `e` (the "root"). If this is a language object, the function in `w$call` is called; otherwise, the function in `w$leaf` is called. To make this into a tree walk, the `w$call` function should at some point recall `walkCode()` for each element in its `e` argument.

As an application to illustrate a tree walk, let's convert an arbitrary language object to an object with the same structure, but with each node being an R `"list"` object, at each level. Notice that just coercing the object to a list as we did above is not the same computation. This only converts the top level:

```
> ee <- as.list(quote(y-mean(y)))
> ee[[3]]; typeof(ee[[3]])
mean(y)
[1] "language"
```

To convert all elements, we need to use a codewalker that calls `walkCode()` recursively for each element and makes a list of the results. Here's a function, `callAsList()`, that does it, along with its own code walker, `.toListW`:

```
.toListW  <- codetools::makeCodeWalker(
    call = function(e, w)
      lapply(e,
          function(ee) codetools::walkCode(ee, w)),
    leaf = function(e, w)
        e
    )


callAsList <- function(expr)
    codetools::walkCode(expr, .toListW)
```

The `makeCodeWalker()` function in **codetools** just constructs the code walker list (there are some additional arguments, which are not needed for our example).

The **base** package has a somewhat related function, `rapply()`, that applies a function to all the non-list elements of a list, at all levels. It is implemented in C, avoiding recursive R calls and so being somewhat faster, particularly if the list structure is deep, with many levels. It is also somewhat simpler, in that the user only specifies what to do with leaf elements, and there is no need to construct a code walker object. And, as with other `apply()`-type functions, it will pass down additional arguments to the function supplied, via its `...` argument.

One catch is that `rapply()` only operates on lists, not on language objects. In order to use it, the language object needs to be converted to a list with the same structure. Which happens to be just what our `callAsList()` function does. Of course, since `callAsList()` uses the R recursive version of the tree walk, there will not be much if any speedup unless `rapply()` is used a number of times on the same object. Another catch is that the internal code calling the user function will not protect the leaf expression from being evaluated; you need to use `substitute()` to get the language object itself.

An example, handled each way, will illustrate. Suppose we wanted to make a table of the number of occurrences of each distinct leaf node. For table purposes, it's best to convert leaf objects to character strings, by calling `deparse()`.

First, with `rapply()`. Its standard mode of computation will substitute every leaf node in the argument with the corresponding value returned by the function supplied, and then `unlist()` the result, producing a character vector in this example. Calling `table()` on this vector is what we want. Here's a function, `codeCountA()` that uses `rapply()` to produce the desired result:

```
codeCountA <- function(expr) {
    ee <- callAsList(expr)
    table(rapply(ee,
        function(x) deparse(x)))
}
```

For the code walking approach, we start with nearly the same code walker as in `callAsList()`, just calling `deparse()` in the `leaf` function. The call to `walkCode()` gives us a list, which we `unlist()` at all levels and pass to `table()`:

```
.countW  <- makeCodeWalker(
    call = function(e, w)
        lapply(e, function(ee) walkCode(ee, w)),
    leaf = function(e, w)
        deparse(e)
    )

codeCountW <- function(expr)
    table(unlist(
        walkCode(expr, .countW),
        recursive = TRUE))
```

You can experiment to be convinced that the two implementations produce the same result (the functions are in package XRexamples). For example:

```
> expr <- quote(y - mean(y))
> codeCountA(expr);codeCountW(expr)
```

```
-  mean    y
1     1    2

-  mean    y
1     1    2
```

Typically we will have some further restrictions on the leaf nodes (such as, only wanting names or only certain names). In either version, our function can test accordingly and return `NULL` if the test fails. The `NULL` elements are dropped by `unlist()`.

# 5.4 Interfaces and Primitives

Everything that happens in R is a function call, but some function calls cause things to happen outside of R, particularly through C code that is part of the R process. Computations outside of R are important for extending R, both for future extensions using such software and for analyzing existing software. An analysis to understand the functional programming validity of a non-trivial computation, for example, nearly always needs to assess some underlying non-R code being used.

This section surveys interfaces in base R and their implications for programming individual functions, especially for understanding how existing interface computations may affect them. If you are planning to add a substantial interface to subroutines in any of these languages in your effort to extend R, you should look at Chapter 16. In particular, I think that the Rcpp interface to C++ provides a substantially more helpful approach than the traditional interfaces described in this section. Even if the non-R software of interest is not itself in C++, the ease and flexibility of the Rcpp facilities may justify a little excursion into that language.

But, for understanding existing interfaces or small-scale programming, the present section should be sufficient.

User-programmable internal interfaces to non-R software all have the form of a call to an R function whose arguments are an identifier for the particular piece of code to be invoked and a sequence of arguments that are to be passed to that code, in an appropriate form:

```
function(.Name, ...)
```

R comes with four such interface functions, although we will emphasize one of them as the appropriate choice for most applications.

R inherited three interface functions from S:

```
.Call(.Name, ...)
.C(.Name, ...)
.Fortran(.Name, ...)
```

(All the functions have some optional named arguments as well, which we can ignore for this discussion.)

For all three interfaces, `.Name` is nominally the character string name of an entry point in the compiled code, either as interpreted by C (for `.Call()` and `.C()`) or Fortran (for `.Fortran()`). In practice, explicit interface calls inside a package will usually replace the name of the entry point with an object generated by the R process that represents the "registered" version of the entry point. The registered version is both more reliable and (somewhat) more efficient (see Section 7.3). It can be created simply by the `useDynlib()` directive in the `"NAMESPACE"` file of the package. In all our examples, we'll assume this has been done and the corresponding registered reference is used (e.g., in the call to `.Fortran()` on page 87).

The `.Call()` interface is the most important for extending R. It takes an arbitrary number of arguments corresponding to R objects and returns a correspondingly arbitrary R object as its value. At the C level, "R object" means a pointer to the structure in C that represents all R objects. It's the responsibility of the interfaced C code to interpret the arguments appropriately for the particular application. More advanced interfaces, such as the Rcpp interface to C++, automate much of the checking and coercion required.

The `.C()` and `.Fortran()` interfaces also, nominally, take an arbitrary sequence of R objects as arguments. However, the corresponding code in C or Fortran does not have an R-specific data type. The code is intended to be R-independent, or at least to be written without needing to know how to manipulate R objects. For this purpose, the data types for all arguments must come from a specified set of types in the compiled language, and the R function calling the interface is responsible for coercing the arguments to the correct type. The limitations imposed on their arguments make these interfaces less satisfactory for most projects that extend R. However, they have been used extensively in the past, justifying a brief look here. For details, see the *Writing R Extensions* manual or [11, Chapter 11].

The `.C()` and `.Fortran()` interfaces differ from `.Call()` and `.External()` in that they do not return an object as their value. Instead, they use the older procedural ("subroutine") computational model as in Fortran. To return information, the code overwrites one or more of the objects supplied as an argument. The traditional Fortran() approach is that some arguments are input (arrays or scalars) and others are output arrays, elements of which are to be overwritten. The `lsfit()` subroutine in Section 2.2, page 26, is typical. For a more realistic example with R, the routine would be modified to use double precision floating point, the internal type used by "`numeric`" vectors in R.

```
subroutine lsfit(x, n, p, y, coef, resid)
double precision x(n, p), y(n), coef(p), resid(n)
integer n, p
```

The arguments `coef` and `resid` are output arrays into which the results will be copied. The R function calling this routine through the `.Fortran()` interface is responsible for allocating all the arrays and passing the corresponding arguments, always with the correct types.

```
lsfit <- function(x, y) {
    d <- dim(x); n <- d[[1]]; p <- d[[2]]
    coef <- matrix(0., p, p)
    resid <- numeric(n)
    z <- .Fortran(lsfitRef, x, n, p, y, c = coef, r = resid)
    list(coef = z$c, resid = z$r)
}
```

Here we used the fact that function `dim()` returns a vector of type "`integer`". R rarely leaves integers unconverted if any arithmetic is done on them, so one needs to be careful, coercing explicitly in any doubtful case. Also, remember that constants such as `1` are *not* integers; an integer constant must have the letter capital "`L`" after it, as in "`1L`". Single precision real data can be constructed and passed to a Fortran routine if necessary; see `?.Fortran` for details.

In addition to these three interface functions, R evolved with its own model for interfacing to C-level code. The most important part of this model is the concept of *primitive* functions, which appear to be ordinary R functions but are in fact special types of object, which we will look at in Section 6.1, page 100. The evaluator deals with calls to primitives by branching more or less directly to some corresponding C code for the specific function, rather than by the general function-call mechanism used for all actual function objects.

The use of primitive functions for low-level computations, such as arithmetic and mathematical computations, reflects the ⌐INTERFACE⌐ principle. These computations are handed over to C-level code, rather than being compiled into machine instructions. The behavior of these primitive interfaces is important for ensuring functional behavior and for studies of computational efficiency. Functional methods can be defined for most primitive functions, treating the primitive as a standard generic function.

The set of primitives is fixed, however, and not relevant as a tool for extending R, so first let's complete our discussion of interface functions. There are two additional interface functions in the R model:

```
.External(.Name, ...)
.Internal(call)
```

The `.Internal()` interface, like primitives, is restricted to use in base R and not available for extensions. A table compiled into R gives the names and properties of primitives and functions available via `.Internal()`. Unlike primitives, the latter functions correspond to actual R function objects. They can do other computations in addition to the C code. But like primitives, this interface is not available for extending R.

The call to `.External()` interface, on the other hand, has the same form and interpretation as `.Call()`, taking arbitrary objects as arguments and returning one as its value. It is an alternative to `.Call()`, differing rather radically in the form of C programming required (based, in fact, on the Lisp model more than on C). Some examples and hints for programming with this can be found in the *Writing R Extensions* manual. The main advantage of `.External()` is in handling functions with arbitrarily many arguments (the "`...`" equivalent), but this can be mimicked in `.Call()` by passing `list(...)` as a single argument.

## 5.5   Getting it to Run Faster

In most of the book, our attention is on extending R by providing something new: a new area of application, the results of research or a new computation of interest.

In some situations, getting an existing computation to run faster, *seriously* faster, is also an extension. A really substantial speedup of a critical computation can extend that computation to a wider range of data.

If this is the goal, the ‾INTERFACE‾ principle and the analysis of function calls in R in Section 3.2 imply that the most promising approach is an implementation outside of standard R software.

R is not designed for all primitive calculations to be done in the language, nor was the original S language. In particular, basic arithmetic and logic are defined for vectors and implemented in R by an interface to "primitive" functions. The set of primitives is fixed but other interfaces are supported to essentially all languages likely to provide high-quality efficient computations.

Interfaces are Part IV of the book. Chapter 12 has a table of interfaces to a number of languages; Chapter 16 discusses subroutine interfaces, often the natural candidate for a speedup; Chapter 15 shows an interface to Julia, a language with some R-like features that aims for low-level efficiency.

Some aspects of interfaces, as discussed in Part IV, that are particularly relevant when interfaces are used for speedup:

- For efficiency and often for convenience, *proxy objects* may be important: objects in R that are simply proxies for objects in the server language. Good interface software lets the user manipulate those objects and obtain information from them naturally in R without the overhead of conversion.

- Also related to data conversion: The natural form for a substantial data set in the server language may not be a good match directly for R. Consider a paradigm where each language takes the data from a common database format but interprets that data in its own natural way. An example in Chapter 14 concerns data in XML files.

- A relevant form of interface not usually called one is to process R language code with a different evaluation scheme. An example is described in [33]. Such specialized interfaces can only be helpful if the R computations can be evaluated with special constraints or assumptions.

For greater efficiency without using an interface:

- The evaluation of given R code may be improved, without changing the code. The R code in the package can always be byte-compiled, using a compiler written by Luke Tierney [35], and included with R.

  Several current projects propose alternative evaluators to that supplied with R, intended to work with all or most R functions.

- Recasting some R computations will allow parallel processing in R , using the **parallel** package for example.

- A long-utilized programming technique in R and S is *vectorizing*, meaning to find alternative R computations that typically replace an iterative computation with one that carries out the same basic operations but replaces the iteration with a fixed number of calls. If the calls are mostly to primitive functions, a substantial reduction in compute time can result.

  The example below illustrates the technique; see also [11, Section 6.4].

All these techniques can be helpful and important for various applications. My opinion, though, is that speedups sufficiently dramatic to constitute an extension of R will more likely come from the interface techniques.

In the example below, timings give a rough estimate that the overhead of an R function call is on the order of $10^3$ instructions. Vectorization gave us an order-of-magnitude improvement over a naive implementation. These numbers are very rough and dependent on the particular example, and are likely to be a substantial overestimate of improvements obtainable in a realistic example. Techniques using the full R evaluation process will be hard pressed to achieve order-of-magnitude improvements in most cases, barring special hardware for parallel solutions or unusually suitable examples in both that case and for vectorizing.

### Example: convolution

The remainder of this section presents an example that illustrates differences in time taken for various versions of a particular computation.

The example is presented in the *Writing R Extensions* manual as an obvious candidate for an interface to C. We require a function to compute the "discrete convolution" of two vectors. If $x$ and $y$ are two such vectors, then their convolution, in the version used by the *Writing R Extensions* manual, is a vector $z$ defined by:

$$z_{k+1} = \sum_{i-j==k} x_i * y_j \tag{5.1}$$

In practice, if one cared to do this computation quickly, the Fast Fourier Transform is the way (see `?convolve` in R and the reference there). But it makes a simple example, which we will revisit in comparing various interfaces to C and C++.

The computation is obviously iterative. Each pair of elements from $x$ and $y$ contribute to one element of $z$, so the natural computation is a double iteration. In a straightforward C formation, if `nx` and `ny` are the lengths of the vectors and if the vectors are converted to pointers-to-double, `xp`, `yp` and `zp`, the whole computation is a simple double loop, assuming `z` has been allocated of length `nx * ny - 1` and initialized to 0:

```
for(i = 0; i < nx; i++)
  for(j = 0; j < ny; j++)
    z[i + j] += x[i] * y[j];
```

The manual uses the example to illustrate interfaces to C, and we will see that others, generally preferable, are made available using a C++interface (Chapter 16).

The manual cites convolution as a computation that is easy to program in C but difficult in R. More precisely, it's not easy to program it *efficiently* in R. R has looping capabilities, so the C code translates directly into R, producing a function something like the following:

```
convolveSlow <- function(x, y) {
    nx <- length(x); ny <- length(y)
    z <- numeric(nx + ny - 1)
    for(i in seq(length = nx)) {
```

```
        xi <- x[[i]]
        for(j in seq(length = ny)) {
            ij <- i+j-1
            z[[ij]] <- z[[ij]] + xi * y[[j]]
        }
    }
    z
}
```

This is a silly computation, in the sense that the individual R operations are working on single numbers. The arithmetic operations are implemented as primitive functions, which will help some. Nevertheless, we would expect the time taken to be large and to go almost entirely into the overhead of setting up the individual calls and managing the R objects involved.

We can run a simple measurement to calibrate this overhead. The number of R function calls will be roughly proportional to the product of the lengths of the arguments, so let's fit a line and plot the times (taking the sum of process and system time from the values returned by the `system.time()` function).

```
> sizes <- seq(100,800,50)
> times <- numeric(length(sizes))
> data <- runif(800)
> for(i in seq_along(sizes)) {
+     x <- data[seq(length=sizes[[i]])]
+     thistime <- system.time(convolveSlow(x, x))
+     times[[i]] <- sum(thistime[1:2])
+ }
> plot(sizes^2, times, type = "b")
> coef(lm(times ~ I(sizes^2)))
  (Intercept)   I(sizes^2)
2.053661e-02 4.690823e-06
```

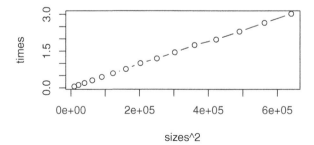

Indeed, the computer time is closely proportional to the number of steps in the inner loop, and the suggestion is that each step takes some number of microseconds. This was run on a machine with a 3Ghz speed, suggesting very roughly $O(10^3)$ computations each time.

Precise estimates here are likely a waste of time: One can argue that each step in the loop is several function calls, but on the other hand the functions are all primitive, involving substantially less work than for ordinary R functions.

A rough but useful number to remember is that the evaluator overhead starts around $10^3$ computations per call, going up as the work involved in setting up the call increases (as discussed in Section 3.2, page 51).

We can compile the function using the R byte compiler, and rerun the same test.

```
> convolveComp <- cmpfun(convolveSlow)
> for(i in seq_along(sizes)) {
+     x <- data[seq(length=sizes[[i]])]
+     thistime <- system.time(convolveComp(x, x))
+     times[[i]] <- sum(thistime[1:2])
+ }
> plot(sizes^2, times, type = "b")
> coef(lm(times ~ I(sizes^2)))
  (Intercept)   I(sizes^2)
7.053505e-03 7.971097e-07
```

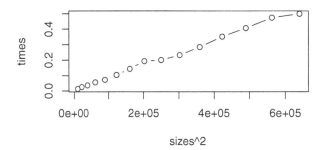

The proportionality constant estimated in this case is over 5 times smaller, a big win for compiling. But remember that this example is extreme, doing a trivial amount of calculation on each call to a primitive function. You will not be likely to achieve such improvements in more realistic software. In the next two versions, compiling makes essentially no difference because the important overhead is no longer in a large number of R function calls. Nevertheless, byte compiling is a good practice once a package is reasonably stable or likely to be used extensively.

The convolution example was introduced in the R manual as a natural for an interface to C or C++. For comparison, however, it can also be treated by vectorizing in R itself, although a little ingenuity is needed. The goal is to reduce the number of calls to R functions. As in many examples, the key is to see that the numerical result corresponds to some whole-object computations using functions already implemented by interface to low-level code, typically C or Fortran.

Notice from the definition that the various elements of the result are sums of pairwise products of the elements of the two vectors. R can form all the pairwise products via the function `outer()`. If these were laid out in an array of the right shape, sums on the rows (or columns) would give the desired result. These sums are also available as efficient functions, for example `rowSums()`. Deciding what is the "right shape" takes a little experimenting: Give it a try if you like before looking below. For more details on various vectorizing techniques, see [11, Section 6.4].

A vectorized version of the computation is:

```
## Convolve computations vectorized as row sums
## of the elementwise products, padded with rows of 0
## and then reshaped so row sums coorrespond to convolution
convolveV <- function(x, y) {
    nx <- length(x); ny <- length(y)
    xy <- rbind(outer(x,y),
                matrix(0, ny, ny))
    nxy <- nx + ny -1
    length(xy) <- nxy * ny # so it fits ...
    dim(xy) <- c(nxy, ny)
    rowSums(xy)
}
```

We can carry out a similar experiment to the one on `convolveSlow()`. In order to see a similar quadratic dependence on problem size, we need to increase the size of the input data, here taking the vectors 10 times larger than before:

```
> sizes <- seq(1000,8000,500)
> times <- numeric(length(sizes))
> data <- runif(800)
> for(i in seq_along(sizes)) {
+     x <- data[seq(length=sizes[[i]])]
+     thistime <- system.time(convolveV(x, x))
+     times[[i]] <- sum(thistime[1:2])
+ }
```

```
> plot(sizes^2, times, type = "b")
> coef(lm(times ~ I(sizes^2)))
  (Intercept)      I(sizes^2)
-1.449029e+00   4.109791e-07
```

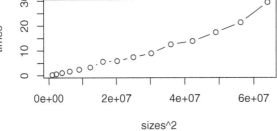

The vectorized version seems to have reduced the constant of proportionality by about an order of magnitude. The range of problem sizes over which this improvement can be expected is limited, however. The reduction in the number of R calls is accompanied by creating larger intermediate objects. As the size of the problem grows, memory requirements will become the dominant limit—even in the example here, there is some super-quadratic growth suggested on the right side of the plot. When I gave students in an advanced statistical programming course an assignment to vectorize this calculation, there were a number of approaches, some of them similar to the above and others partially vectorizing the computations (with a single loop rather than a nested one). The resulting timings varied quite widely, with some iterative versions that reduced the number of calls to linear with the size showing competitive improvements to the fully vectorized version above.

Compiling this function had essentially no effect on the timings, further evidence that in the vectorized version other computational overhead than function calls is responsible for most of the computing time.

For the third approach, we can apply a C version of the computation, as was the point of the example in the *Writing R Extensions* manual:

```
SEXP convolve3(SEXP a, SEXP b, SEXP ab)
{
    int i, j, na, nb, nab;
    double *xa, *xb, *xab;

    na = LENGTH(a); nb = LENGTH(b); nab = na + nb - 1;
    xa = NUMERIC_POINTER(a); xb = NUMERIC_POINTER(b);
    xab = NUMERIC_POINTER(ab);
    for(i = 0; i < nab; i++) xab[i] = 0.0;
```

```
    for(i = 0; i < na; i++)
        for(j = 0; j < nb; j++) xab[i + j] += xa[i] * xb[j];
    return(ab);
}
```

Not to worry if you are unfamiliar with C programming: The essential point here is that the computation does essentially the same iteration as our naive R implementation. But now the iteration is done in a language where "function calls" do much less computation, and support a much more elementary paradigm.

For a timing comparison, we will use an R function `convolve3()` that uses the C implementation via the `.Call()` interface:

```
> sizes <- seq(1000,8000,500)
> times <- numeric(length(sizes))
> data <- runif(800)
> for(i in seq_along(sizes)) {
+     x <- data[seq(length=sizes[[i]])]
+     thistime <- system.time(convolve3(x, x))
+     times[[i]] <- sum(thistime[1:2])
+ }
> plot(sizes^2, times, type = "b")
> coef(lm(times ~ I(sizes^2)))
  (Intercept)    I(sizes^2)
-2.055817e-04  1.276478e-09
```

In this version the proportionality estimate is reduced by over two orders of magnitude relative to the naive R function, supporting the general rule of thumb that R function calls cost on the order of $10^3$ basic computations.

Again, the important general lesson is that *if* one needs to obtain substantial speedup, then either a major shift in the underlying algorithm or a direct interface to a lower-overhead language (or both) will likely be required.

# Chapter 6

# Objects

This chapter discusses some topics about objects in R that may be helpful in planning and implementing extensions. If everything that exists is an object, what does that mean about their internal structure (Section 6.1)? What resources do objects need and how are they managed (Section 6.2)? Since environments are essential for object references in R, how should one think about environments *as* objects (Section 6.3)?

## 6.1   Object Structure: Types and Attributes

The ⌈OBJECT⌉ principle ("everything is an object") applies quite literally at the implementation level, as becomes clear if we examine the C-level code underlying R. In the implementation in C of the `convolve()` computation shown on page 94, the arguments and return value all have the same C type:

```
SEXP convolve3(SEXP a, SEXP b, SEXP ab)
```

The type `SEXP` is a (pointer to) a single structure defined in the internal C code for R. All objects in R share a compatible internal representation as this structure. Some details of the structure are documented in Chapter 1 of the R internals manual. For extending R, the key point is that objects in R are *self-describing*; that is, all the information about a particular object is determined by looking in the object itself.

Local computations may make assumptions about what particular type of data the structure represents, as the code on page 94 does in treating the arguments as numeric vectors. But those are always assumptions, which may need to be verified dynamically. Contrast, for example, the Java language, which has objects and classes but where internal variables may also be primitive, non-object types. Information about such data comes from the program, in the form of type declarations, not from the data itself.

At the R level as well as in the .Call() interface to C, it is perfectly legal, and even sometimes sensible, to allow a totally arbitrary object in computations. In object-oriented contexts, this corresponds to the class "ANY", literally matching any R object.

The C structure is composed of a variety of internal fields. Two fields are important in programming: the type and the attributes. The attribute field is in effect a pointer to a named list of properties; the type field holds one of a fixed set of integer codes that distinguish objects for further computation. The concepts of attributes and types go back to the S software on which R is based (types were called modes in S).

For most programming purposes, one needs to assume some specialization of the objects involved. With R, the distinction should usually be based on the object's *class*. Through object-oriented programming, classes provide a well-defined and extensible hierarchy describing the content and computational facilities for objects. An object's class is the value of class(x), corresponding to the "class" attribute of the object or a predefined value for basic objects.

The type field is used for internal computations and by the functions in the base package implemented in C. For extending R, classes are the relevant way to organize objects. For understanding basic computations, the implications of types may be important. They determine how remaining fields in the internal representation are organized. Also, being a C-level enumeration, types lend themselves to quick checks and switch-style organization of alternatives. The type of an object is returned as a string by typeof(x). The full discussion of types and other internal structure is in the *R Language Definition* manual at the r-project.org web site. The official list of possible types is in Section 2 of the manual.

The *vector* types are:

```
[1] "logical"   "double"   "character"  "complex"
[5] "integer"   "raw"      "expression" "list"
```

These implement the concept taken from S of vectors as objects that can be indexed numerically or logically. All these objects share essentially the same internal structure, with a pointer to a block of memory containing the values for the individual elements in the vector. The XRtools package has a function typeAndClass() that takes any number of arguments and returns a matrix with the corresponding classes and types. For a few of the vector types:

```
> typeAndClass(1.0, 1L, TRUE, "abc", 1i)
```

```
                1              1L          TRUE        "abc"        0+1i
      Class "numeric" "integer" "logical" "character" "complex"
      Type  "double"  "integer" "logical" "character" "complex"
```

The classes and data types for vectors have the same names, except that data type `"double"` corresponds to class `"numeric"`.

The internal structure supports natural and efficient primitive implementations of the `` `[ `` and `` `[[ `` operators for vector types. The distinction of vector types is confused somewhat in that base R allows some indexing operations on some non-vectors; for example, single elements of `"language"` objects by numerical index.

Of the non-vector types, the most relevant for extending R is `"environment"`; we will look at it in Section 6.3. Function calls and function definitions as objects are also not vectors, although they intuitively have "elements". They can be manipulated in vector-like ways by coercing them into `"list"` objects. Section 5.3 shows some examples. Similar C computations applied directly to these objects require quite a different style. Internally, the objects are organized as Lisp-style lists; some instructions and related macros can be found in the *Writing R Extensions* manual, but programming at this level is usually something to avoid.

Function call objects all have type `"language"`. Intuitively, language objects should be thought of as any object that does not evaluate to itself—things that are not "ordinary" objects to the evaluator. There is a slight confusion in base R on this point. Name objects are language objects by this definition and according to the base function `is.language()`, but they have a different type, `"symbol"`. To deal consistently with language objects use either the virtual class `"language"` or `is.language()`, not the type.

Objects representing an expression need to be created by a special kind of function call, to avoid the expression being evaluated. Functions `quote()` and `substitute()` are the most often used. If it is easier to construct the character string form, then `parse()` will construct language objects, but beware that it always produces a vector of type `expression()`, whose first element is typically the language object you wanted. For name objects only, the function `as.name()` can be used.

The classes for objects of type `"language"` distinguish them according to their role in the R grammar, as we saw on page 52:

```
> typeAndClass(quote({1;2}),quote(if(x > 3) 1 else 0),
+              quote(x * 10))
       quote({    quote(if (.. quote(x * ..
Class "{"        "if"         "call"
Type  "language" "language"   "language"
```

Function definition objects are all of class `"function"` but can have three types.

```
> typeAndClass(function(x)x+1, sin, `+`, `if`)
      function(x) x .. sin          +           if
Class "function"       "function" "function" "function"
Type  "closure"        "builtin"  "builtin"  "special"
```

Ordinary functions, defined in R itself, have type `"closure"`. When R was first implemented, the decision was made to implement some basic functions specially, directly in C. These are the "primitive" functions, all in the **base** package and not available for extension by users. Two additional types, `"builtin"` and `"special"`, apply to these functions. The types differ in whether the arguments in the function call are evaluated before the internal code is called ("yes" and "no", respectively).

Computations to extend R are quite unlikely to work with the primitive function types as objects. These are only allowed in the **base** package and do not have the properties such as environment, formal arguments or body that would be used for computations with the objects. If a computation needs to distinguish primitive functions, use the function `is.primitive()` to test the object.

Other types are either highly specialized or hidden from the user.

- Type `"S4"` was added to the language to handle objects from formal classes, when those classes do not inherit from one of the normal types. We will deal with the distinction when discussing the definition of classes in Chapter 10. From a programming perspective, the `"S4"` type is an implementation mechanism. Objects from subclasses of an R type should behave correctly for that type, while objects that do not inherit from a type should correctly refuse to behave as if they did.

- Type `"externalptr"` is for "pointers" generated in some non-R way; R essentially guarantees to leave them alone. So should we. The objects are useful in some interface situations; in particular, the Rcpp package provides some tools to handle them and to relate them to particular C++ classes (Section 16.5, page 354) and the rJava interface uses them to handle general Java objects (Section 12.6, page 251).

- Type `"NULL"` is just for the NULL object. As a class, it is often useful to mark special cases or undefined elements which might otherwise be arbitrary objects from a given class. (See Section 10.5, page 170 for an example.)

- Type `"pairlist"` is for Lisp-style lists; they were part of the original implementation of R and still arise in various internal tasks, but are not part of the programming interface. Types `"language"` and `"closure"` for language objects and function definitions use the same internal structure. See the *Writing R Extensions* manual for details.

- The types `"promise"`, `"char"`, `"..."`, `"any"`, `"bytecode"` and `"weakref"` are for internal use only. They will never be needed for programming at the R level (well, hardly ever).

## Attributes

Other than type, the most important built-in property of the object structure allows the definition of *attributes*. These are properties, each a subsidiary object specified by name, for example `attr(x, "dim")`. An attribute's value may in principle be any object other than `NULL`: the `attr()` function returns `NULL` for any attribute not found. Attributes can be assigned as usual by using the corresponding replacement function version of `attr()`:

```
attr(x, "dim") <- c(3,4)
```

Assigning `NULL` to the attribute deletes it.

Some attributes have special meaning in base R; `"dim"` is an example. Assigning the attribute as above turns `x` implicitly into a matrix, but with no `"class"` attribute. The numeric vector on the right side is converted to `"integer"`. When extending R, these special treatments may need to be made more consistent, as is true with S3 classes (Section 10.8). Defining an equivalent formal class or a subclass to regularize the attributes is often a good idea.

All types of objects can potentially have attributes but they should not be used for environments or other reference types, as discussed in Section 6.3.

The collection of non-null attributes of an object are traditionally referred to in S and R as the *attribute list*, and they can be obtained in this form as `attributes(x)`[1]. This is a list with names. As in many examples of such objects in R, the vector nature of the list is probably best ignored: Dealing with the attributes only by name and not by "position" in the list avoids bad things happening if attributes are added or deleted.

The attribute list implements a very old and key concept in S and R: Structural information can be included in an object in a form independent of the specific data, or of its type. An advantage of separating structure from data relates to the ⟨INTERFACE⟩ principle. The form of storage in R for numeric matrices, for example, matches and derives from the similar form in the important Fortran subroutines

---

[1]But unfortunately, this returns `NULL`, not the empty list, if there are none.

for matrix computations. Keeping structural information separate from the data in objects from the `"matrix"` class retains compatibility with these computations.

In general, much of the data important to statistics, science and information management has the form of basic data—typically a vector in our terminology—supplemented by structural or contextual information (dimensions of an array, levels of a factor, etc.). Separating the attribute list from the data itself in the fundamental structure for R objects is therefore a useful general mechanism. Computations with the attributes can largely be done independently of the type of data; similarly for slots in formal classes, which are implemented as attributes.

The notion of vectors with structure is made explicit in functional OOP by the class `"structure"`. The use and usefulness of attributes for non-vector types is more variable. Language and function definition objects have a different internal organization. They can still have attributes and slots, but one complication has to do with the `"names"` attribute. Names for elements of a vector, for named arguments in a function call and for formal argument names in a function definition are all subsumed in the use of the function `names()` to extract or replace these. However, the implementation of names differs. Vector names are an attribute, but names in the other two cases are stored internally.

```
> z <- list(a=1, b=2)
> names(z)
[1] "a" "b"
> zExpr <- quote(list(a=1, b=2))
> names(zExpr)
[1] ""  "a" "b"
> ## But:
> attributes(zExpr)
NULL
```

Stick to the function `names()` or the general function `attr()`, which takes account of the different implementation.

# 6.2  Object Management

R is implemented as a C program, so memory management uses the C-style storage mechanisms in a variety of ways. But the $\boxed{\text{OBJECT}}$ principle implies that we mainly care about the management of R objects, and that is conceptually a simpler question. Basically, R objects are allocated when needed and de-allocated sometime after they are no longer referenced. The "sometime" is when the R process does an explicit garbage collection.

When are objects needed and therefore allocated? In two cases mainly:

1. when some R code or some C interface code called from R needs a new object; or

2. when an object must be duplicated because it's about to be modified and there may be more than one reference to it.

The second case, object duplication, is of interest if one is trying to economize on memory usage. For large objects, we may at the least want to understand when the objects are being copied, to help in improving computational efficiency. A debugging tool exists for this explicit purpose: `tracemem()`, which prints an informative message when a specified object is duplicated. See Section 5.2, page 79, for an example and comments. It is useful to look at the actual implementation of object duplication, but first let's consider the overall strategy for object management.

Object allocation implements the dynamic nature of objects in R, inherited from the original S software: objects are created when the user does a computation that explicitly needs that object. They are not declared in a programmatic sense.

Ensuring that objects are not overwritten incorrectly and that the storage in them will eventually be retrieved requires keeping track of object references. R only de-allocates objects, as noted, when the process decides that it is time to compact storage, by a process of "garbage collection". This may occur at any time an allocation is requested, and so is not predictable, for an R user or programmer, although it can be forced by a call to the R function `gc()`, which can be a useful debugging technique; for example, with errors possibly related to overwriting memory.

When garbage collection takes place, all objects that are not required by the evaluator in its current state will be de-allocated. The evaluator knows that it needs a particular object if it is referenced in certain environments or found on an internal list of protected objects. The environments include those associated with packages that have been loaded into the R process and the frames for each of the function calls currently active. Special lists are used in C level programming to protect objects that are not assigned in R but need to stay around at the C level, temporarily or permanently. (The `PROTECT()` macro calls seen in C code for R computations have this purpose.)

Open-source systems, such as R, by definition allow one to examine just how computations are implemented. That can involve reading some convoluted, long and not-too-clearly-commented C code. But it's a key advantage of open-source software; namely, the source *is* open and therefore available to check what is really going on. This is at times a major advantage.

For a fundamental internal action such as object duplication in R, the effort may be worth it. First, one has to locate the relevant code. In this case, the key fact is that all copying to protect locality is done by the C routine `duplicate()`.

One could find this out from several sources: discussions on the R mailing lists, working down in the R source itself and the online R internals or language definition manuals (in this case, in the *Writing R Extensions* manual).

The code for duplicate() is found in file "duplicate.c" in the "src/main" folder of the R source. The actual computation is done in a local subroutine, duplicate1(). This routine is essentially a C switch statement on the internal type of the data. In the listing below, I have simplified the code by leaving out cases that are very similar and by omitting some details of the coding that are crucial but not directly relevant to duplicating. I've also added comments that give the character string name of the type corresponding to the C-level symbolic integer, e.g., "environment" for ENVSXP.

After the code, we'll analyse the computations. They give some useful insight into the basics of different object types in R.

```
static SEXP duplicate1(SEXP s, Rboolean deep)
{
    SEXP t; R_xlen_t i, n;

    switch (TYPEOF(s)) {
    case ENVSXP: /*type "environment" */
        /* ... and a number of other reference types */
        /* ... do nothing */
        return s;
    case CLOSXP: /* function definitions */
        /* ... some special stuff */
        break;
    case LANGSXP: /* type "language" */
        /* ... special duplicate for pairlist */
        t = duplicate_list(s, deep);
        SET_TYPEOF(t, LANGSXP);
        DUPLICATE_ATTRIB(t, s, deep);
        break;
        /* ... and similarly for other pairlist types*/
    case VECSXP:  /* type "list"  */
        n = XLENGTH(s);
        t = allocVector(TYPEOF(s), n);
        for(i = 0 ; i < n ; i++)
            SET_VECTOR_ELT(t, i,
              duplicate_child(VECTOR_ELT(s, i), deep));
        DUPLICATE_ATTRIB(t, s, deep);
        break;
```

```
        case LGLSXP: /* "logical" vectors */
        DUPLICATE_ATOMIC_VECTOR(int, LOGICAL, t, s, deep); break;
            /* ... and similarly for "integer", "real", etc.*/
        case S4SXP:   /*  S4 classes that do NOT contain other types */
            t = allocS4Object();
            DUPLICATE_ATTRIB(t, s, deep);
            break;
        default: /* and an error message for unknown type */
            UNIMPLEMENTED_TYPE("duplicate", s);
        }
        return t;
    }
```

First in the listing come some types that are reference objects, environments and other types. Precisely because they are references, nothing is done. Section 6.3 discusses computations for these.

All the other data types are duplicated. Notice that in each case, a separate C computation duplicates the attributes of the object, in exactly the same way each time. Here again is the principle that attributes are independent of the type of the object.

The types representing Lisp-type lists are used for language and function definition objects mainly (Section 5.3) and use a special form of duplication for those. Then there are the vector types. Type "list" is duplicated recursively, as you would expect: A new list vector is allocated and all its elements are assigned with the duplicates of the original elements.

Atomic vectors use special, simpler code via a C macro. Finally, S4 objects that are of type S4 rather than of another type essentially only have attributes (called slots in these objects), so it's only necessary to generate a simple S4 object.

The C routine duplicate1() has a second argument, deep. Copying of a list in any language can be shallow (copy only at the top level) or deep (copy the list and all its elements, recursively). The actual duplicate() operation itself is always a deep copy.

# 6.3 Reference Objects; Environments

The fundamental reference in R is the combination of an environment and the character-string name of some object in that environment. For programming, this means that the software provides a name and the environment is either inferred from the context or supplied explicitly. The evaluator finds a reference matching a name by searching the local frame and its parents. An expression such as `stats::rnorm` is an explicit reference to the object named `"rnorm"` in the environment of exports from package `stats`.

Most R objects are *not* themselves references: changes to an object in the frame of a function call are local and do not alter any external object, either an argument or an external reference—the essence of functional programming in R.

But some R objects must be references; in particular, environments themselves are R objects, of type and class `"environment"`. A change made to an environment is reflected everywhere that environment object is being used.

Environments are essential for the evaluation of all R computations, but usually in the background, with no need for the user to be concerned. They are essential, in addition, for the approach to encapsulated OOP discussed in Chapter 11, but here also the bulk of the programming can proceed without explicit acknowledgment of the implementation in terms of environments.

One can compute directly with environments in R, without the more abstract notions of OOP. A variety of R functions deal directly with these objects, mostly by analogy to named lists, with functions and operators programmed to behave in a similar way with environments. From a programming viewpoint, this approach has some dangers.

As a general rule one should never assign attributes to an environment or any other reference object, unless you are generating the environment for specific purposes. Even then, it's better to take the extra step of defining a functional OOP class that extends `"environment"`, allowing slots to be assigned in a normal way without corrupting the environment.

More subtle dangers arise because both the `` `$` `` and `` `[[` `` operators will extract or replace named elements in an environment, apparently as they would with a named list object:

```
x$a; x[["b"]]
x$flag <- TRUE
for(n in theseNames)
    x[[n]] <- NULL
```

If the computations come in a function that could take either a named list or an environment as argument `x`, the result of assignments will be different in the two cases.

Because environments are reference objects and lists are not, the external effect of the replacement will be different. A computation with no side-effects when `x` is a list will have side-effects when it is an environment.

Also, the behavior of the computations themselves can be inconsistent between lists and environments. The `` `$` `` operator applies partial matching to lists but not to environments: if object `x` had exactly two elements, `"aa"` and `"bb"`, the expression x$a will return the "aa" element if x is a named list but NULL if x is an environment. Conversely, assigning NULL to a named element of a list removes the element, but for an environment stores the explicit value.

A function that does these computations and that is being provided for general use should avoid such possibly dangerous ambiguities. Usually the function should be specialized explicitly to one or the other case and should check and warn, at least, if the argument is not of that class.

A substantial amount of pre-OOP and S3 class software uses lists with named elements as an informal kind of class structure. The most wide-spread example is software for fitting statistical models. The S3 class "lm" for linear models returns a fitted model object that is a named list with elements "coef" and "resid" for the coefficients and residuals.

It would be dangerous and unreasonable to pass reference objects to functions written for such objects. The function might well modify the named elements in the belief that the changes would be local to the function call. A computation like

```
z$resid[i] <- epsilon
```

would cause unintended side effects if z was either an environment or an object from an encapsulated OOP class.

Here the onus is on the person writing the calling function to avoid using a reference object as the argument to an existing function that expects a named list.

But suppose you are writing your own functions or using other functions that are *intended* for reference objects. You expect the function to update the object passed as an argument.

Even here, some caution is needed in the form of the call. Basically, the reference object should only appear once in the call, so that in effect there is no ambiguity about which argument involves that object.

The issue in this case is lazy evaluation. The actual arguments in an R function call may not be evaluated immediately. In particular, it is possible that changes made from outside by using a reference object will take place before some arguments are evaluated. For a simple example, suppose function updateScore() does some computation to revise an element named "Score" in an object x, with a second argument, value, to updateScore() being used for some purpose, it doesn't matter what. The dangerous form of call would be something like:

```
updateScore(x, x$Score)
```

Now suppose that updateScore() does some test on the data to decide whether to zero out the Score component before updating it:

```
updateScore <- function(data, value) {
```

```
    if(zeroFirst(data))
        data$Score <- 0
    data$Score <- data$Score + value
    data
}
```

The nature of the test is irrelevant but notice that it does not involve `value`, meaning that lazy evaluation of this argument is delayed until after the test and, perhaps, zeroing the score and changing the value supplied in the call. (Keep in mind that in practice, there may be a page or so of other code between the optional zeroing and the increment of the score.)

The computation is potentially incorrect, in what I consider the worst way: silently returning a wrong result, in this case by carrying out the computation with 0 instead of the intended `value`.

Lazy evaluation is a useful technique and in any case has been part of S and R for so long that revising it would be an unacceptable break with compatibility. However, it puts constraints on computations with reference objects. In this example, the rule of "Only one occurrence of the object in the call" would require something like:

```
value <- x$Score
updateScore(x, value)
```

The assignment will now be unambiguously done before the function call.

Occasionally it's important to recognize general reference objects (perhaps to warn about using them in just such a functional context). A class union, `"refObject"`, exists for this purpose. The expression

```
is(x, "refObject")
```

tests generally for an object which behaves according to reference semantics.

Reference objects are available for extending R, usually through the definition of reference classes (Chapter 11). Proxy objects in R for objects in other languages, using one of the XR interfaces (Chapter 13), will also be reference objects in this sense. If you defined a special class with reference semantics in some other way, that class could also contain `"refObject"`.

# Chapter 7

# Packages

## 7.1 Understanding Packages

Functions are the natural unit for programming in the small: creating software that answers questions of immediate interest and that captures specific ideas in extending R.

Packages are equally the natural unit to capture the next level, when the new ideas or the new application starts to have some substance and particularly when the extensions may be useful to others. They are the way to combine related ideas about new computations (functions), new data structures (classes) and potentially software from other languages (interfaces). Packages build naturally on other extensions to R when these are also organized as packages. R packages have a structure that encourages documentation. Repositories such as CRAN and mechanisms such as github encourage sharing packages with others.

The present chapter examines the role of packages for extending R. We will not cover all the basics of creating a package. Any moderately advanced book or online guide to R should include some guidance; for example, [11, Chapter 4]. Interactive development environments for R such as RStudio (www.rstudio.com) often provide tools for creating packages. The book *R Packages* by Hadley Wickham [37] gives a step-by-step guide using the RStudio interface.

### Package structure

The initial version of R inherited from S a functional, object-based language, meaning that the bulk of programming would consist of function definitions. In S, no particular structure was assumed for the corresponding source code.

Some projects would also use the subroutine interfaces to C and Fortran; these subroutines would also exist in both source form and as object code, most often collected into dynamically linked library files, in the UNIX operating system style. Again, the organization of the original source code was essentially arbitrary.

One of the first and certainly one of the most important extensions developed for the initial release of R was to organize the R functions and possible subroutine interfaces into the *package* structure. Collections of software for R would be organized in a directory structure in which the R and subroutine source code would be found in files within specific subdirectories.

Additional structure was added, then and during later evolution, to include documentation, data, tools and miscellaneous other files. Following the UNIX paradigm, file names would have suffixes that identified the source language or other role of the file within the package.

From our perspective of creating extensions to R, a package is a directory with a structure that R can interpret as source files for this package. Within the conventions imposed by this structure, a package can contain essentially any software or data that the project wants to organize and provide to users. We'll refer to this directory as the *source directory* for the package.

The structure required is prescribed in detail in Section 1.1 of the *Writing R Extensions* manual. Three subdirectories are most important for extending R:

- "R", containing R source files;

- "src", containing compilable source, principally C, C++ or Fortran along with some special files related to these; and

- "inst", for sending an arbitrary collection of files and subdirectories to the installed package.

The original conceptual model for the "R" directory was a collection of files containing function definitions. These source files would be evaluated when the package was installed, but usually the only effect would be to create the corresponding function objects.

The R source for modern packages may do more than just define functions. Both object-oriented programming and interfaces to other software need to do additional computations.

Object-oriented programming in R includes function calls, from files in the "R" subdirectory, that are evaluated when the source package is installed; for example, to setClass() and setMethod(), and to similar functions for encapsulated OOP. Side effects of these calls create objects in the installed version of the package that will in turn be used when the package is loaded to make class and method definitions available.

The interface packages of Chapters 14 and 15 allow functions and classes to be created in an application package that are proxies for server language analogues. The computations to create these proxies will also be specified from files in the "R" directory.

The contents of the "src" directory are all related to interfaces. The traditional structure is described in the *Writing R Extensions* manual. The subroutine interface described in Chapter 16 uses tools to automate some of the requirements for the "src" directory.

The "inst" directory will be used in the interface packages of Chapters 14 and 15 to provide server language code required by the package.

To understand how the structure of the package directory affects an application to extend R, we need to look at the process by which the source package is made available for users.

## Package processing

R defines a process for packages to transform the source directory so it can be used in an R session. There are two steps: *installation* and *loading*. The computations required are themselves mostly done by R functions, allowing us to examine and experiment.

Installation is a command line computation that transforms the source directory into another directory, which we'll refer to as the *installed* version of the package.[1] The contents of the source directory are processed in various ways to prepare the package for use. The installation step should be preceded by a command that builds an archive file from the source directory. Installing from the archive leaves the source directory unmodified.

Invoked directly from a terminal application, these are carried out by R commands build and INSTALL.

```
R CMD build directory
R CMD INSTALL archive
```

where *directory* is the source directory and *archive* is the archive file created by build. For a package XRexamples:

```
$  R CMD build XRexamples
*  checking for file 'XRexamples/DESCRIPTION' ... OK
    .... (more messages) ....
*  building 'XRexamples_0.5.tar.gz'
```

---

[1]Sometimes called the *binary* version, but it can contain more than binary (compiled) files.

```
$  R CMD INSTALL XRexamples_0.5.tar.gz
* installing to library '/Users/jmc/Rlib'
* installing *source* package 'XRexamples' ...
    .... (more messages) ....
* DONE (XRexamples)
```

We'll look at what happens in Section 7.2.

Loading is an R function call, to `loadNamespace()`, that uses the installed package from the library to create a namespace environment and integrate that with the current R process:

```
> loadNamespace("XRexamples")
<environment: namespace:XRexamples>
```

Loading also creates a second environment, named `"package:XRexamples"` in this case, containing the exported objects from the namespace. Only these can be imported or accessed by the `` `::` `` operator. This is the environment inserted into the search list if the package is attached, as it will be if `loadNamespace()` is called via `require()` or `library()`. (See Section 3.1, page 49.)

Section 7.3 examines the load step.

## Package setup step

Most of the techniques discussed in this book fit automatically in the install-load process, but some will require extra steps either when loading the package or before installation. We'll look at load actions in Section 7.3, page 119.

A number of current development tools for R packages, including some described in this book, go beyond the installation-and-load paradigm in that they may create files in the source package. The files created may be in the `"R"`, `"src"` or `"man"` directories, for example. The tools are frequently R function calls themselves, but cannot be part of the installation procedure since they create files that this procedure needs.

These computations are not part of formal package development as seen in the *Writing R Extensions* manual. My suggestion for projects extending R is to make them an explicit part of the package, however, as a *package setup step* that would precede installation. Recipients of the package would not be required to run the setup step, but seeing it explicitly clarifies the package contents by documenting the setup computations. Also, in circumstances where a package is being developed cooperatively, as is possible with Github for example, all developers will be able to run a consistent setup if that needs to be revised.

Standard installation recognizes a `"tools"` subdirectory in which one or more setup scripts could be included. The `build` command includes the `"tools"` directory but it is does not alter installation and is not copied to the installed version of the package. A reasonable approach to formalizing the setup step is to include one or more source files in this directory that carry out the setup.

Package XR has a function, `packageSetup()`, to run such scripts. From an R session, one calls `packageSetup()` with the file(s) supplied as an argument. Called with no arguments, `packageSetup()` looks for a file `"setup.R"` in the `"tools"` directory. The convention is that the evaluation takes place with the working directory set to the package's source; `packageSetup()` takes an optional argument for this directory and checks that the location is indeed a package source. It also provides a few extra steps that may occasionally be needed. If a version of the package has already been installed, the evaluation takes place in an environment with the package's namespace as its parent, meaning that the functions and other objects used in the script do not need to be exported.

Further discussion of load actions and package setup is in Section 13.7, page 283. The example of a setup script shown there is for an interface, but the use of `packageSetup` would be essentially the same for other uses.

One example arises with the Rcpp interface to C++. That package has an R function, `compileAttributes()`, that creates both R and C functions to define the interface to a particular C++ function, inferred directly from the C++ source code. This is a substantial improvement over writing the corresponding functions by hand. The catch is that the call to `compileAttributes()` cannot be part of the regular R source in the package, since it will generate code in the `"src"` directory too late in the installation procedure.

Applications using an interface to other languages, such as those in Chapters 14 and 15 may create proxy functions or classes in R. These computations often use metadata information in the server language to create the proxy object. Very desirable, but it does imply that the evaluator for that language is available when the R package is installed. If the package is being installed by a repository such as CRAN, this may be a problem for some interfaces.

An example of a setup step with proxy functions and classes for a Python interface is shown in Section 13.7, page 285.

As a different example, the roxygen2 package [38] provides for R documentation in the form of formatted comments in the source code. This has advantages in keeping source and its documentation in sync as well as being generally easier to write than the "raw" documentation files. (The XR packages discussed in Part IV use this form of documentation.) It too requires the use of an R function, in this case `roxygenize()`, to generate target files in the `"man"` directory of the package.

## 7.2   Installing a Package

Understanding what the `INSTALL` utility does helps clarify what preparations are needed in the source package.

We're assuming here that the installation is from an archive file created by applying the `build` command to the source directory. In this case, `INSTALL` begins by expanding the archive file into a new directory, referred to as the *source directory* below.

`INSTALL` uses the contents of the source directory to create a corresponding *installation directory*—a subdirectory of the library directory with the name of the package. Various further files and subdirectories are created under the installation directory.

The essential steps in installation are actually carried out by R functions in the **tools** package. Here are the main steps. To check for yourself or see more details, get a source version of R and examine the local function `do_install_source()` in `"tools/R/install.R"`.

1. The `"DESCRIPTION"` file is parsed, producing a named list of parameters for installation, each corresponding to a directive such as `Imports:` in the file.

2. A check is made that all packages named in the `Imports:` and `Depends:` directives are available.

3. If there is a currently installed version of the package, it is moved to a backup.

4. If the source directory has a `"configure"` file, that script is executed.

5. A modified `"DESCRIPTION"` file is generated for the installation directory. Other standard files, such as `"NAMESPACE"`, are copied from the source directory.

6. If the source directory has a subdirectory `"src"`, the files in this directory are compiled and installed as a dynamically linked (shared) library file under the `"libs"` subdirectory of the installation directory.

7. The R source files in subdirectory `"R"` of the source directory are concatenated into a single file of R source (plus an assignment of the package name). A `Collate:` field in the `"DESCRIPTION"` file is used if one exists to determine the order of the files. The output file is parsed to check its correctness.

8. Data files in an optional `"data"` subdirectory are processed into a database for "lazy" loading.

9. Additional files and directories are copied into the installed package. The complete file system tree of the `"inst"` subdirectory of the source directory is copied to the installation directory; so, for example, file `"inst/abc"` is copied to file `"abc"` in the installation directory.

10. If directories `"demo"`, `"tests"` or `"exec"` exist in the source, these are copied to the installation directory. (The directory `"tools"`, however, is not copied to the installation, although it will be picked up by the `build` command.)

11. The file of R source code created from files in the `"R"` directory is evaluated with two extra steps, a special "lazy loading" mechanism and an optional byte-compilation of R function objects; the result is an R "database" file in a specialized format that will be used to load the package.

12. R detailed documentation files from the `"man"` subdirectory of the source directory are processed to produce a database file similar to that for the source, but in this case for documentation. Some auxiliary files are also created, all in the `"help"` subdirectory of the installation directory.

13. Optional package "vignettes" are processed and index files for documentation, vignettes and demos are produced.

14. Information from the package `"NAMESPACE"` file is parsed and saved.

15. A separate R process is run to test that the installed package can be loaded.

The description above is fairly detailed, although the implementation (mostly in file `"install.R"` of base package tools) is much more complicated than this description. It's worth referring to the details to resolve what's actually happening when not all goes as expected.

Three steps directly relate to the design of packages to extend R:

- Contents of the `"src"` subdirectory are compiled and linked in step 6, with output in the `"libs"` subdirectory;

- The files and subdirectories of the `"inst"` subdirectory are copied directly to the installation directory in step 9.

- The R source in the `"R"` subdirectory is evaluated in step 11;

Step 6 carries out a fairly standard compilation and linking for C-compatible software, based on the process for compiling R itself. It can be customized, provided you have some familiarity with the programming tools used: see the *Writing R Extensions* manual.

Step 11 provides great flexibility via R itself. The original paradigm mainly expected function definitions to be assigned at this time, but other R computations can be done as well, as we considered on page 110.

The "inst" directory is the miscellaneous bin for R packages. The installation and load procedures make no assumptions about what's in that directory. At step 9, the contents are copied as is into the library directory for this package. They are available both at load time and when the package is running in the R process. The package's location in the library is available from function system.file(), which allows package software to make use of all files and subdirectories copied into the installation directory.

## 7.3   Loading and Attaching a Package

The installation procedure creates a subdirectory corresponding to the package name in the library directory chosen for the installation. Loading the package into an R session makes functions, compiled software and other objects or files in the package available to other packages and to the expressions in the evaluator. Traditionally, this was done by calling the function library(), supplying that function with the name of the package to be attached. The original paradigm was that the objects in the package would then be read in (essentially by evaluating the source code for the package) and made available as one of the environments in the search list (as shown on page 123).

The procedure has become more general and the potential contents of the package have become more varied. There are now two distinct steps:

- The package is *loaded* into the session, creating and registering a corresponding namespace environment. The contents of the package are now available, including dynamically linked compiled code if any, but only by explicit reference to the package name.

- The loaded package may be *attached* to the search list, by inserting into the search list a package environment containing the objects exported by the "NAMESPACE" file.

Packages can be loaded without attaching them to the search list. In particular, if one package imports from another, the second package need not be attached unless users are expected to call functions from that package explicitly.

The package loading procedure is carried out by loadNamespace(). Called directly or from another function that requires the package to be loaded, it constructs a namespace environment for the package and takes actions to load related information into the R process. Here is a list of the main steps.

1. Information about the package that was saved during installation, including that from the `"DESCRIPTION"` file, is retrieved and checked to verify that this package can be loaded.

2. A namespace environment is created and registered for the session, corresponding to the package name. The parent of that environment is also a new environment, which will contain the imports for the package. The XRtools package has some functions to retrieve information from the namespace, such as the packages from which it imports.

3. The objects specified in the `import` directives of the `"NAMESPACE"` file are found and assigned in the parent environment of the namespace. This includes assigning the imported class objects and the generic functions corresponding to imported methods.

4. The R code object created at step 11 of the installation procedure is sourced in the namespace environment.

5. Data objects are loaded according to `"sysdata"` and `"LazyData"` mechanisms, as described in the *Writing R Extensions* manual.

6. S3 methods are registered.

7. Any dynamically linked libraries of compiled code included by `useDynLib()` directives in the package's `"NAMESPACE"` are loaded into the process (see page 120).

8. The `.onLoad()` function for the package is called, if there was one.

9. Information about classes and methods in the package is cached in corresponding tables for use in the session. For details, see Section 10.4, page 161.

10. Any load actions registered by the package source code at install time are carried out (see page 119).

11. The exports from the package are assigned in the exports environment (i.e., the package as it will appear when attached). The exports environment for package myPackage is given the name `"package:myPackage"`.

12. The namespace is sealed, so that no ordinary assignments will be allowed.

13. Any "user hooks" defined for the package are evaluated.

This is the essential procedure that controls what information can be used in the package. We consider next a few steps that are important in making your package behave as intended.

## Package imports

Packages should import explicitly all the functions and other objects they use from other packages, except for the **base** package which by definition must always be available. Imports will be specified by directives `import()` or `importFrom()` in the `"NAMESPACE"` file. The information from these directives is stored internally in the installed version of the package. At step 3 during loading, all the objects implied by the import directives will be assigned in a new environment. This becomes the parent environment of the package's namespace; the parent of the imports environment is the namespace of the **base** package (see Section 3.1, page 51 for an example).

Importing explicitly is good housekeeping in that it makes clear where all the references in the package are expected to come from. Traditionally, package authors tended not to bother with imports if that package would "obviously" be attached; either it was in the default search list or the expected context implied that it would have been attached in the user's session. That saved some effort. Also, `"NAMESPACE"` files have not always been required for packages. They were introduced into R by Luke Tierney [34] and are now an essential part of the package concept.

Current practice strongly suggests being explicit about all imports and some repositories, including CRAN, require it. Given that, the simple solution is

```
import(thisPackage)
```

for all packages used. This imports all the objects in the package, wasting some space for large packages but not enough to worry about.

A stronger argument for naming the individual objects is to be precise and clear about what comes from where. It's possible that functions with the same name exist in more than one of the imported packages. The `importFrom()` directive gives all the information and selects each object from the intended package. The catch is that any change to the list of functions requires editing the `"NAMESPACE"` file.

A function in the XRtools package, `makeImports()`, will output any missing `importFrom()` directives in the current version of a package. For example, the SoDA package provided tools and examples for my 2008 book on R, [11]. The version installed on CRAN at that time was casual about importing from the default packages. Running `makeImports()` on that package:

```
> makeImports("SoDA")
## Found, but not in package: .Random.seed
importFrom(methods, new, is, metaNameUndo, `packageSlot<-`, ....
importFrom(stats, as.formula, terms, resid, update, runif)
importFrom(utils, data, menu, recover)
```

The printed `importFrom()` directives can be added to the package's `NAMESPACE()` file. With a long list of imports from a package such as the imports from methods in the example, one may prefer the neater if less explicit option of `import()` for the entire package.

## Load actions

A package may need to take some specialized steps at load time. For example, information that depends on dynamically linking compiled code is not available until the corresponding step in loading the package. On the other hand, any computation that needs to store information in the package's namespace cannot take place after loading, since the namespace is then sealed.

A flexible way to provide such customizations is via load actions, specified during installation and carried out at load time, in step 10 above. R provides tools, in the methods package, to create a list of such actions during installation. The function `setLoadActions()`, for example, takes any number of functions as its arguments and arranges to call these at load time; `evalOnLoad()` takes an expression object and arranges to evaluate the expression at load time. In both cases, the typical technique is to construct the function or expression during installation to carry out the particular computation needed using information available at load time.

A motivating example is a mechanism in the Rcpp package to provide information about C++ classes so the class can be used from R. The R version of the class needs a pointer to the C++ class to invoke methods for that class. The pointer will be set when the C++ code is linked, so it cannot be created during installation. On the other hand, looking up the pointer dynamically every time it is needed would be extremely inefficient. The solution is a load action: the Rcpp software that creates the R version of the class constructs a function object designed to find and assign the pointer for this class. The function is passed in a call to `setLoadAction()` during installation. At load time, each of these functions will be called with the side effect of assigning the desired pointer in the namespace of the package that uses this class.

Notice that the action will be set in the application package that creates the class, but the writer of that package does not need to know anything about the required action, the details are all handled by the Rcpp package.

Information like a C-style pointer is only defined from the running R process. Some other specialized information needed by the package may require R computations incorporated into the package, but which is available before loading. If the information is used to create files in the package's source directory, it *must* be produced before. These computations can be carried out in a package setup step as discussed on page 112. The dual possibilities of load actions and setup steps will be examined again in the context of interface packages; see Section 13.7, page 283.

Older mechanisms for load actions are unsatisfactory: They are either too restricted (`.onLoad()`) or take place too late to store information in the namespace (user hooks). See the online documentation, `?setLoadActions`, for more details on load actions.

## Load actions and setup steps

Load actions and the execution of a script in a setup step (page 112) are both mechanisms to include information in an application package that may not be available when the package is installed; for example, if the installation is taking place in a central site in order to distribute the installed version of the package.

Load actions and setup steps are alternatives for many situations. Sometimes only one applies. For example, the function `compileAttributes()` in theRcpp interface to C++ writes both R and C++ source code into the application package to define proxies for C++ functions (Section 16.3, page 343). The resulting code has to be compiled during installation before the package can be loaded, so only a setup step can use `compileAttributes()`. In contrast, only a load action can produce external pointers as used in the C++ proxy class example above.

When both apply, load actions have the advantage of not requiring a separate step. There is a non-technical detail: The `INSTALL` command does a "test load" by default. Delaying something to a load action doesn't help unless the repository takes the step of running the command with the `--no-test-load` option.

The setup step, on the other hand, generates explicit R source that can be modified, for example, by adding roxygen-style documentation. (But the code needs to be edited or regenerated if the relevant server language facts change, such as the fields in a class.)

## Linking compiled subroutines

R package installation includes an optional step to compile source files in the `"src"` subdirectory of the package source directory. The language for the source code is inferred from the file suffix; as this is written the possibilities include C, C++,

Objective C, and various standard versions of Fortran. Package installation then creates a "shared object" file containing the corresponding object code.

The shared object file is saved in the `"libs"` subdirectory of the installation directory. Step 7 in the load process will link any such objects available in an available package, if the `"NAMESPACE"` file includes a directive `useDynLib()` with the name of the package as its argument:

```
useDynLib(XRcppExamples)
```

for example, would make available the shared object created by installing package XRcppExamples. If you create the package via the `Rcpp.package.skeleton()` function, the package's own shared object is added to the `"NAMESPACE"` file automatically. (And you must have the directive; the load process for a package does not automatically link its own shared object.)

Additional arguments to the `useDynLib()` directive specify entry points in C or Fortran that will be called from R. This mechanism can be useful if you are writing the `.Call()` interface directly or using one of the other basic interfaces such as `.Fortran()`. Just supply the name of the routine to be called. The installation and load steps will create an R object of class `"NativeSymbol"`, and assign it in the package's namespace with the same name. The object contains the information needed to identify the entry point when the corresponding shared object file is linked into the R process. This is only relevant if you are not using the Rcpp package's `compileAttributes()` mechanism discussed in Section 16.3, page 343. That mechanism generates a subroutine and corresponding character string argument for `.Call()`.

Suppose we define directly the `pointerString()` function to use the `.Call()` interface to call the C routine `pointer_string`. The C code for `pointer_string` is in the `"src"` directory of the source package; therefore, the corresponding C entry point will be compiled into the library's shared object. If this had been the only entry point needed, the `"NAMESPACE"` file for cppExamples would have the line:

```
useDynlib(cppExamples, pointer_string)
```

and the R code would provide this name as the first argument to `.Call()`:

```
pointerString <- function(x)
    .Call(pointer_string, x)
```

referring to the object `pointer_string` of S3 class `"NativeSymbol"` created by the `useDynlib()` directive.

The same mechanism is used for Fortran interfaced directly through `.Fortran()`. For example, the SoDA package calls the subroutine `geodistv()` through `".Fortran"`. The modern version of the package has, in the `"NAMESPACE"` file:

```
useDynLib(SoDa, geodistv)
```

and the corresponding `.Fortran()` call has the form:

```
res <- .Fortran(geodistv, ....)
```

The object created in this case is from the S3 subclass `"FortranRoutine"` of `"NativeSymbol"`, allowing possible differences in Fortran entry points to be handled.

The `useDynLib()` directive can access shared object files other than the package's own and can take symbol arguments in those cases as well.

If you happen to have an entry point and an R object with the same name (typically, you gave the function calling a routine the same name as the routine), you can rename the entry point when declaring it. If there had been a `geodistv` R object already in package SoDA, changing the directive to

```
useDynLib(SoDA, geodistv_sym = geodistv)
```

would allow using `geodistv_sym` as the argument to `.Fortran()`.

## Search list

If the package is attached, via a call to `require()` or `library()`, the export environment created at step 11 of the load process is inserted into the "search list", with the result that exported objects are available to the R evaluator by simple name, assuming no environment earlier on the list has an object of the same name.

The search list is not a list in R but a sequence of environments, each environment being the parent of the preceding environment. The sequence is presented to users as a character vector of environment names.

If we invoked R in the usual way and then attached the package Matrix, the available packages plus the global environment and a few miscellanies would be shown by a call to `search()`

```
> search()
 [1] ".GlobalEnv"          "package:Matrix"
 [3] "package:lattice"     "package:stats"
 [5] "package:graphics"    "package:grDevices"
 [7] "package:utils"       "package:datasets"
 [9] "package:methods"     "Autoloads"
[11] "package:base"
```

This character vector of names is just a way of representing a sequence of environments, one per package, such that the parent of one package's environment is the environment for the next package. As an example:

```
> ev <- as.environment("package:stats")
> ev
<environment: package:stats>
attr(,"name")
[1] "package:stats"
attr(,"path")
[1] "/Users/jmc/R_Devel/r_devel/library/stats"
> identical(parent.env(ev),
+           as.environment("package:graphics"))
[1] TRUE
```

The last expression is TRUE because, in our current session, package graphics follows package stats in the search list.

# 7.4 Sharing your Package

Sharing the new capabilities in your package is probably the most constructive step in programming with R. The vast and diverse set of packages in collected repositories and other sources is an enormously valuable resource. Over the years, easy availability of so many techniques has been revolutionary in the opportunities to analyse data or build on research in statistics and other data-based disciplines.

The growth in R packages has benefitted from and shared in the internet revolution. Back in the early days of S, sharing software meant distributing a physical medium (typically magnetic tape). Even the first released version of R was at least nominally preserved on compact disc (page 36). Part of the revolutionary impact comes from the simplicity of downloading and installing packages.

Your package *is* the source directory. The simplest approach is to put this in some publicly accessible place and announce it. For downloading, the archive file produced by the `build` command would be the most convenient.

A number of projects from the R community and others have developed facilities for sharing software that take some further steps, which are worth considering as soon as you want to share packages:

- Sharing the installed version of the package (called the "binary" version by CRAN) in addition to or instead of the source version would simplify installation for those using the package, particularly if installation of the package requires compilation. This does require that the installed package distributed be compatible with the operating system type and perhaps version on which the package is to be used.

- In the other direction, packages that are still evolving may want to encourage suggestions for changes or to have the development process be shared by a team working from the same master copy of the package. For this purpose, the source package should be supported by a good version control system. The package will be shared by copying (or "cloning") the source and its version control information.

As this book is written, there are a number of helpful projects that have produced distribution mechanisms reflecting these and other ideas, and new projects are proposed or underway. We will look here at just two, admittedly the two most active now, but others offer variations or additional facilities and by the time you read this there may be important new possibilities.

The mechanism closest to the central R development itself is the CRAN repository. It is also arguably the most trusted and accepted by users, although as its documentation makes clear, no distribution site can feasibly validate the computational quality of the packages distributed. For users of the repository, CRAN provides:

- an archive file for the package source;

- when possible, "binary" archive files, that is, archives of the installed directory for the package on various platforms;

- separate files of documentation, including a collected version of the detailed documentation and versions of any "vignettes" supplied with the package;

- a variety of other information including results of running the `check` command, previous versions of the source archive and dependency information from and to this package.

The archive files here are condensed `tar` files for source and suitable condensed archive files for the target operating system for the binary versions.

For the recipient, CRAN provides source for the package (as an archive file to be expanded) and much more. For the author, the submission and approval process may require some changes to the package not otherwise planned, although many of the requirements amount to validating the package along lines that would be wise anyway.

There are also constraints on the contents of the package (laid out on the CRAN `Policy` link), reasonable and usually necessary to maintain the repository.

As the range of extensions to R expands, the procedures for a particular repository may be difficult to match with the needs of some packages.

One difficulty arises with packages whose operation depends on software not always available with R. From the viewpoint of the repository maintainers, it's not practical to maintain arbitrary non-R software for all potential uses. At the same time, CRAN would like to verify that packages submitted are at least nominally usable, and so requires that the package installs and loads.

Packages using a connected interface in the terminology used in Part IV, for example, may not be designed to install without the target server language software being available. These packages are likely to have initialization requirements that need the server language itself and possibly some nonstandard modules as well. One possible approach for application packages using such an interface to install in a more restricted situation will be shown in Section 13.7, page 283.

Even with the constraints required by the repository, the thousands of packages on CRAN and its continued growth are evidence that the benefits of distribution and visibility make the effort worthwhile for many projects.

As a contrast, but also widely used by R developers, GitHub follows a "shared development" model, in which the "publicly accessible place" for the package is provided centrally under the site `github.com`. The remote directory for a particular package is typically a mirror of the source directory. But it is also a "repository" in the sense of source-code development systems; that is, it contains revision information for successive revisions of the package software. Specifically, GitHub uses the Git revision control system originally developed for Linux. There is nothing specific to R packages in the basic mechanism, and indeed GitHub is widely used for many kinds of software projects.

A potential user of the package clones a copy of the complete source directory. The cloned directory is then used in `build` and `INSTALL` commands like any other package source. The revision control makes it simple to update at any time to incorporate changes; in fact, it's a feature of GitHub that the cloned directory is itself a repository with all the revision and history information. The owner of the central repository, who will have established an account with GitHub, will upload revisions that users will pick up by synchronizing their copy. There is also a mechanism for more advanced users to "fork" a version of the package, make suggested changes there and then communicate those to the owner of the package (by a "pull" request).

This model can be extended to give multiple users permission to update the central repository. GitHub supports creation of "organization" accounts, in which a specified set of users are team members. One or more repositories will then be moved under the organization, effectively as subdirectories. Team members are allowed to make revisions to the central repositories.

GitHub has become widely used; as of some time ago there were over 10 million "repositories", including many R projects.

The two approaches appear highly contrasted, but in fact can be usefully combined. To avoid the conflicting use of the word "repository", let's refer to the CRAN-style approach as *package distribution* and the GitHub-style approach as *package exporting*.

The package exporting approach is designed for projects in a continuing state of evolution. Aside from having permission to update, everyone using the site is essentially operating on the same level: cloning the original material, in our case the source directory for a package, and then installing the package locally just as for a package the user wrote personally.

One could adopt this approach at any stage in a project, and arguably sooner is better than later to get the benefit of others' advice. (Personal preference will vary here.)

Package distribution services, in contrast, provide a great deal more for the ordinary user; most importantly, perhaps, some feeling that the package has gone through a reasonable amount of inspection and testing, although again it's simply not feasible to perform substantive testing of arbitrary submissions, beyond those provided with the package itself. Users at least know that the package was in fact installed on some platforms, those for which binary versions are available. Binary versions are particularly helpful if the package requires compilation that may depend on particular languages or compilers for them. Separate downloads of documentation also allow prospective users to get some feeling for the package's capabilities before installing it.

From the author's perspective, the package distribution approach is valuable for communicating software to a wider audience, once the software is in a reasonably well-developed state. Revisions will often continue and the package distribution mechanism works well for communicating those to users, but each revision needs to be as operational as the last. In contrast, package exporting will typically put out each significant change one at a time, and ideas that sounded good but turn out to be unfortunate are perfectly OK, and to be expected.

Combining the two approaches is attractive, with an exported version that starts earlier and evolves more frequently along with a distributed version that reflects user-friendly stages. Many R-related projects now utilize this dual approach, either specifically with GitHub and CRAN, or with variants of one or the other.

The interface, tools and examples packages cited in the book are available from `https://github.com/johnmchambers`

# Chapter 8

# In the Large

The small (functions) and medium (packages) scales of programming discussed in Chapters 5 to 7 are software-driven, for immediate use or as software that now has recognized longer-term value. R has an intrinsic structure that makes functions and function calls the way to approach specific computations and a large set of programming tools has evolved to make packages the natural way to organize non-trivial collections of functions and other tools.

But extending R is often part of a larger project, not the goal itself. Realistic applications of computing usually ask how software can be designed to respond effectively to the needs of the application.

What kind of data are we dealing with? What properties of that data need to be available for computing?

What computational results do we need, once the data are available? In particular, what are the computations for which good methods may be challenging to implement? Is there existing software that provides some or all of the required computations?

By their nature, questions like this are unlikely to have neat, universal answers.

There are some useful approaches, however, that can be described. The rest of this book is largely devoted to two of them. Object-oriented programming, in whatever form, rests on the idea that the structure of data can be described more or less formally in terms of classes of objects, with computations organized as methods related to these classes. The key advantages of OOP in these terms are that it localizes the information about data and computations, and that generalizations to extend the software can utilize inheritance to build on the existing base cleanly. Part III of the book will discuss the R version, including questions of designing OOP structure for extending R.

A key concept for programming in the large as envisioned here is that one should look for the best available computation for an important task. If this turns out to be in a form other than standard R functions and packages, consider using that form, with as convenient an interface to R as possible.

Modern computing is too broad and includes too many good implementations to restrict one's choice when tackling a serious application. For many languages general interfaces from R exist. Also, it may be reasonable to add a new interface or extend an existing one, by understanding how interfaces can work. Part IV of the book examines a number of interfaces from R to computations in other languages and to subroutines linked to the R process.

Subroutine interfaces are the original mechanism in R and continue to be an important resource; Chapter 16 discusses a recommended approach. Other languages have become equally important for various computational needs. Chapter 12 describes a variety of possibilities, lists some available interface packages and discusses some requirements for the effective use of interfaces. Chapter 13 presents a uniform, high-level structure applicable to some attractive languages for interfacing. Examples of interfaces in this form, to Python and Julia, are in Chapters 14 and 15.

# Part III
# Object-Oriented Programming

This part of the book explores the ideas of object-oriented programming, from the perspective of extending R. OOP allows extensions that diversify and enrich the underlying data structures for many projects, in a clear and uncluttered form. Chapter 9 presents the basic programming mechanisms for two versions of OOP in R.

The original and most commonly occurring version of object-oriented programming is the paradigm derived from Simula and similar languages. It is what most discussions outside of data analysis and R assume when mentioning object-oriented programming, with no further qualification. Our focus however is on a dual view of OOP, of which this is only one. To make the distinction, we call this paradigm *encapsulated* object-oriented programming. "Encapsulated" is used in computing in a few different senses, one of which we co-opt for our purposes: This version of OOP encapsulates both object properties and methods within the class definition. (More precisely, within the definition of the current class and that of all its superclasses.)

Languages that are based on functional computing, such as R and Julia, are suited to an alternative paradigm, *functional* object-oriented programming. Functional OOP defines methods for functions corresponding to classes for the arguments. Conceptually, method definitions are stored within the function definition. The properties of objects from the class are stored in the class definition, as before.

R supports both functional OOP inherited from S and encapsulated OOP (*reference* classes in R). The two paradigms have both been valuable and successful in computing with data. For extensions of R where the properties of objects are important, one (or both, in some cases) of the paradigms can be extremely helpful. Chapters 10 and 11 discuss functional and encapsulated OOP, respectively. Section 9.2 discusses the choice between the two.

# Chapter 9

# Classes and Methods in R

## 9.1   OOP Software in R

In Chapter 1, we discussed object-oriented programming in both the encapsulated and functional paradigms without actually prescribing how to use them. The present chapter provides the basic answer for R.

Following the ⌷OBJECT⌷ and ⌷FUNCTION⌷ principles, the implementation of an OOP concept as part of extending R consists of two kinds of information. First, a specification of the properties for the objects. Second, the specification of the functions that make things happen with such objects. These are provided by the *class* and *method* definitions.

A class is defined and created in R by a call to `setClass()` for a functional class or to `setRefClass()` for a reference class. These functions both return a generator for the class; that is, a function to be called to generate new objects from the class. They both also have the side-effect of creating a metadata object with the definition of the class. When the class is created by the source code of a package at package installation time, the metadata object becomes part of the package's namespace. Loading the package makes the class definition available for computations in the R process. Section 9.3 discusses creating classes.

Methods are defined in the case of functional OOP by calls to `setMethod()` and in the case of encapsulated OOP by calls to an encapsulated method of the generator object, `$methods()`. These add metadata corresponding to the new methods. In the encapsulated case, the methods become part of the class definition itself. In the functional case, since methods belong to functions rather than to classes, the metadata is accumulated in the package namespace in tables for each generic function. This information also becomes available when the package is loaded into the R process. Section 9.4 presents computations for creating methods.

Section 9.2 considers the two versions of OOP and the choice between them in an application. Section 9.5 presents two additional examples to illustrate the techniques.

This chapter covers the basic computations needed for OOP in R. Special techniques, choices between alternative approaches and an understanding of what happens all require more detailed discussion. Chapters 10 and 11 will provide these, for functional and for encapsulated OOP respectively.

## 9.2   Functional and Encapsulated OOP

Suppose `p` is an object from a class in an encapsulated OOP implementation in some language. Then if the implementation defines, say, a property `sizes` and a method `evolve()` for the class, both are accessed from the object:

```
p$sizes
p$evolve()
```

All the information needed is available from the class definition for `class(p)`.

In contrast, when the software is built on a functional version of OOP, user computations with the object are naturally via function calls. The goal of object-oriented classes and methods is now to generalize commonly occurring functions to deal with an open-ended variety of objects. In a modeling situation, a function

```
update(fit, newData)
```

may be intended to update a fitted model for the inclusion of some new data. Appropriate methods may depend on the class of either argument or both. Encapsulated OOP is not appropriate, because it associates methods with classes rather than with functions. Would one search using `class(fit)` or `class(newData)`?

Implementations of encapsulated OOP may also need methods for functional computation, such as element selection via the `` `[ ]` `` operator. Special mechanisms, sometimes referred to as "operator overloading", provide methods for certain functions and operators while keeping the methods encapsulated in the class definition. Overloading typically uses a reserved set of method names, limiting its applicability. See Section 11.6 for the R situation; here, functional methods can be defined for encapsulated classes.

The distinction implied by the term "functional" may be relevant in another sense as well. Common encapsulated OOP languages treat objects as *mutable references*; that is, methods and other software have the ability to modify the object on which the method is invoked through a reference to that object. Pure functional programming has no such concept. Functional languages such as Haskell tend to have a syntax that does not include procedural modification of objects. Classes and methods built on such a foundation would then be functional as well, and would not support references.

In practice, functional *computing* does not always imply strict functional *programming*. A language that subscribes more or less to the FUNCTION principle (everything that happens is a function call) may not also follow the principle that the function evaluation is local and without side effects. Haskell builds functional programming into the language itself. R encourages functional programming but does not enforce it. Julia has functional OOP but uses object references essentially throughout.

In R, the functional OOP implementation is largely consistent with functional programming: assigning a new value to a slot only changes the object in the current function call. Assigning a new value to a field in a reference class object, on the other hand, has a global effect on that object. While the implementations are consistent, understanding precisely what will happen sometimes requires thinking carefully.

Having a reference object as a slot in a functional class, for example a slot of class `"environment"`, means that changing some data *within* that slot will have a side effect. For this reason, such slots need to handled carefully. A function could create a local version of the whole (non-reference) object but modifying the reference slot would alter the external version of the functional class object.

Conversely, a field in a reference class object will behave according to what it is, unrelated to the reference class object. If you pass the field as an argument to a function, local replacements in that function have no effect on the reference object, nor should they. We will examine these points when discussing the respective functional and encapsulated OOP implementations.

Given the availability of two paradigms, which of the two suits a particular project best? The content of objects from the class can be thought about to a considerable extent without picking a paradigm. The objects will be used quite differently, however.

One simple question about a proposed class often suggests the most natural choice. Suppose we have an object from the new class, and assign a new value to one of the object's properties. Technically, the object remains a member of the class—`class(x)` will not have changed. But how do we regard the new object? In particular:

> If some of the data in the object has changed, is this still the same object?

The answer will be "No" if the object is viewed as the result of functional programming. The object's definition *is* the expression (the function call) that produced it. An object that has been modified no longer represents that functional computation; indeed, strict functional programming languages would usually not contain expressions to "modify an object". (R is not so strict, but a relevant fact is that all replacement expressions in R actually end up assigning a new object to the name, as discussed in Section 5.2.)

Statistical models and other summaries of data are naturally viewed functionally. A least-squares regression model returned by `lm()` will no longer represent that computation if some of its properties (e.g., the coefficients) are modified.

In contrast, objects that have a mutable state are by definition available for modification of that state. The object must have been generated by an initializing method, but its state at that time is usually not its final state, and often no final state is defined or relevant. For example, a model object that was created to simulate a population rather than as a summary of data remains valid as the simulation proceeds. The purpose of a method like `p$evolve()` is precisely to modify the object `p`.

These two examples are straightforward; the paradigm matches the design goals for the class with little ambiguity. That won't always be the case. The same information and essentially the same structure can suit either a functional or encapsulated class. The natural answer to our question above may vary with the purpose of the current computations.

For example, when we examine OOP approaches to the objects known as data frames in R, either paradigm may be preferred depending on the context.

The data frame may represent a data source that evolves through time. Even if the data represents a single time point, it may evolve as we attempt to correct errors in the recording or interpretation. An encapsulated class suits this purpose, such as the `"dataTable"` example on page 136. On the other hand, if the data is part of a functional analysis, that particular object must remain what it is, to have any hope of reproducible results. One would want any tweaking in a functional computation to be local to that function, as would be true for the original `"data.frame"` class.

The important guideline is to be clear about which alternative is being followed in a particular context, so that users' understanding corresponds to what will happen.

After presenting the basic programming steps in the next two sections, we will return to models as an example of contrasting implementations in Section 9.5.

# 9.3   Creating Classes in R

The essentials of a class definition are properties, inheritance and methods. The properties and inheritance of classes are defined in R by calls to functions `setClass()` and `setRefClass()` for functional and encapsulated classes respectively. Methods are specified by separate calls to `setMethod()` or `$methods()` in the two cases. To create the class:

```
setClass(Class, slots=, contains=)
setRefClass(Class, fields=, contains=)
```

`Class` is the character string name for the new class. Both calls return a function object. Calls to that function generate objects from the class. It is usually convenient for users to have the generating function assigned with the same name `Class`. The generator returned by `setRefClass()` also has encapsulated-style methods; in particular, calls to `$methods()` define the encapsulated methods for this class.

In R, properties are called *slots* in functional OOP and *fields* in encapsulated OOP. They are each accessed by name. The class definition specifies a class for each property in the object. Only objects from the specified class or a subclass can be assigned as the corresponding property. A typical value for the `slots=` or `fields=` argument will be a named vector of the desired classes for the properties:

```
fields = c(edits = "list", description = "character")
```

Slots and fields can be left unrestricted by specifying class `"ANY"`.

Properties can be extracted by using the operator `` `@` `` for functional slots or `` `$` `` for reference fields. Using the same operator on the left side of an assignment will set the corresponding property. Properties are also assigned in the call to the generator function for the class. Fields in reference classes may be specified as read-only, in which case only an initial assignment is allowed.

The inheritance patterns among classes are specified by listing the superclasses in the `contains=` argument. The argument will consist of the names of direct superclasses, which must be previously defined and available when this class is defined. Objects from the new class will have all the properties of the superclasses, as well as any properties specified directly.

There are additional arguments to both `setClass()` and `setRefClass()`. Those that are likely to be useful will be discussed in Chapters 10 and 11.

For an example of a functional class, let's consider data such as that collected by tracking devices. Such data arises often from GPS technology, used to record movements of animals and for other similar applications. The devices typically record position and time. They usually have a mechanism to download data in tabular form, such as a `".csv"` file.

A simple class to represent such data might have two slots, `"lat"` and `"long"`, both expected to be vectors of positions in degrees of latitude and longitude, and a slot `"time"` to record the corresponding times. Then the definition of a functional class `"track"` of this form could be as follows:

```
track <- setClass("track",
    slots = list(lat = "degree", long = "degree",
                 time = "DateTime"))
```

The call specifies the names of the properties and requires the first two to be an object of class `"degree"` and the third of class `"DateTime"`. The definitions of those two classes must be available when class `"track"` is created. The complete class definition is in package XRexamples on GitHub.

Objects from the class are generated by calls to the function `track()`. In this example, a likely call would be from a function in an application package that read in data from a file or connection of observations generated by one of the devices:

```
function readTrack(file, ...) {
  ## read a table of data from the file and
  ## extract the xlat, xlong, xtime variables
  ....
  track(lat = degree(xlat), long = degree(xlong)
        time = xtime)
}
```

The generator function for class `"degree"` is likely to validate numeric data as representing latitude or longitude. Specifying the slots to have this class rather than just `"numeric"` suggests that we would like such validation. A class definition can include a method for checking validity. In Section 10.5, page 180, a validity method is shown for the `"degree"` class.

For an example of an encapsulated class, we define a class `"dataTable"`, intended to hold data-frame-like objects with an emphasis on editing and in general "cleaning up" the data. As with the function that constructed a `"track"` object, it's natural to think of this data coming into R from a file, a table in a database, a spreadsheet or other similar common sources for data. But where the previous software created the object, taking its definition "as is" from the file, we are now concerned with the stage of the analysis in which the data is considered and adjusted. In many applications this phase is an important part of the project, one in which R software can participate. A reference class is more natural for this purpose since one expects to make a sequence of changes to the same object, in contrast to creating a new object with each change.

The `"dataTable"` class has as its properties an environment object to contain the variables (the columns) and a vector of names for the rows in the conceptual table:

```
dataTable <- setRefClass("dataTable",
  fields = c(
    data = "environment",
    row.names = "data.frameRowLabels"
  )
)
```

Objects from this class will have fields `"data"` and `"row.names"` from the specified classes. We've taken over the special class `"data.frameRowLabels"` from the `"data.frame"` class; it can be either a numeric or character string vector.

For an example of inheritance, here is a reference class that inherits from `"dataTable"`:

```
dataEdit <- setRefClass("dataEdit",
       fields = c(
         edits = "list",
         description = "character"),
       contains = "dataTable")
```

Objects from class `"dataEdit"` will have the two fields specified directly and also the fields defined for class `"dataTable"`.

The essential implication of "inheritance" in R is that an object from the subclass will be valid when used as an object from the superclass. All the methods defined for the superclass will be available for the new class as well. R has several mechanisms for defining inheritance, but the simple inheritance produced by the `contains=` argument is the most important. Because R enforces consistency of the class of inherited fields or slots, no transformation is needed to pass the object from the subclass to an inherited method. A method for `"dataTable"` can be invoked on an object from class `"dataEdit"` and expect to find suitable objects in the fields `"data"` and `"row.names"`.

Classes in R, both functional and encapsulated, may optionally be created as *virtual* classes. By definition, these are classes from which objects may not be generated. They exist so that other classes can inherit, and share, methods and possibly properties defined for the virtual class. Any class can be defined to be virtual by including the special class `"VIRTUAL"` in the `contains=` argument. The most common form of virtual classes (for functional OOP only) are class unions, which are defined by a call to `setClassUnion()`. Details and examples are in Section 10.5, page 168.

# 9.4   Creating Methods in **R**

Methods are the action part of class definitions. In the case of functional OOP, they are literally the *method* to compute the value of function calls when the arguments in the call match one of the methods defined for this function. In the case of encapsulated OOP, methods are functions contained in the class definition. Calls to these functions are the actions that can be invoked on an object from the class.

## Functional methods

Functional methods are part of the definition of a generic function. They are conceptually stored in the function in a table indexed by the class or classes required for the arguments. Methods are created by a call to `setMethod()`:

```
setMethod(f, signature, definition)
```

where `f` is the generic function or its name and `signature` specifies which classes are required for arguments to the function if this method is to be used.

The `definition` argument is the function object to be used as the method in this case. It must have the same formal arguments as the generic. When the method has been selected, the **R** evaluator will evaluate the body of this function in the frame created for the call to the generic. We will look at the details in Section 10.7.

A frequently useful method for a class is one that automatically prints an object from the class when that object is the result of evaluating an expression. Automatic printing calls the function

```
show(object)
```

from the **methods** package. The signature for the method will specify the class for the argument `object`. The method definition will be a function with the same formal arguments as the generic, in this case just `"object"`.

```
setMethod("show", "track",
          function(object) {
              .....
          })
```

A call to the generic function evaluates the body of the method without rematching arguments (`show()` is called automatically).

Method signatures can include more than one argument in the signature. The generic function `asRObject()` is defined in the **XR** package, and discussed in Chapter 13:

```
> formalArgs(XR::asRObject)
[1] "object"    "evaluator"
```

The XR package defines some methods with just the first argument in the signature:

```
setMethod("asRObject", "vector_R",
          function(object, evaluator) {
              .....
          })
```

This method will be used if argument `object` comes from class `"vector_R"` or a subclass, regardless of the class of the `evaluator` argument. Package XRJulia also has a method definition for `asRObject()`:

```
setMethod("asRObject", c("vector_R","JuliaInterface"),
          function (object, evaluator) {
              value <- callNextMethod()
              if(is.list(value) && doUnlist(object@serverClass))
                  unlist(value)
              else
                  value
          })
```

This method will only be used if `object` matches `"vector_R"` as before *and* argument `evaluator` comes from class `"JuliaInterface"` or a subclass.

When creating a more specialized method, it may be useful to call the previous method and either do some extra computation after or prepare the arguments first. The method above is an example. It begins with a call to `callNextMethod()`, which in this case will call the method with just `"vector_R"` in its signature. The rule is that `callNextMethod()` will call the method that *would* have been selected if the current method had not been defined.

The functions `show()` and `asRObject()` were created as generic functions, just so methods could be written for them. Methods can be defined also for ordinary functions originally written to take arbitrary objects as arguments. It's recommended to precede the first `setMethod()` call for a non-generic function with a call to `setGeneric()`. This in effect declares that the function will now have methods defined for it. A frequent example is the function `plot()` in the graphics package. If our package defined a method to plot an object from class `"track"` the package source would contain:

```
setGeneric("plot")

setMethod("plot", c("track", "missing"),
```

```
function(x, y, ...) {
    . . . . .
})
```

As one starts to define methods for functions with more extensive argument lists, getting the correct formal arguments for the method and the correct version of the signature can be more error-prone. A convenient tool to generate a valid version is `method.skeleton()`. This function writes out some R code that calls `setMethod()` appropriately, given the generic function name and the method signature. If we wanted a method for the operator `` `[` ``, for example, when the object was from class `"track"`:

```
> method.skeleton("[", "track")
Skeleton of method written to [_track.R
```

which would write a file containing:

```
setMethod("[",
    signature(x = "track"),
    function (x, i, j, ..., drop = TRUE)
    {
        stop("need a definition for the method here")
    }
)
```

## Encapsulated methods

Encapsulated methods are encapsulated in the class definition. They can be included in the `setRefClass()` call, but for readability it's better to define them in one or more following calls invoking `$methods()` on the generator.

Methods themselves are functions with arbitrary arguments. The object itself is normally *not* an argument. Instead, fields and other methods for the class are just referenced by name.

Like functional classes, reference classes can have a `$show()` method that will be called for automatic printing when the object is the result of a complete expression. A simple method for the `"dataTable"` class on page 223:

```
dataTable$methods(
    show = function() {
      cat(gettextf("dataTable with %d rows\nVariables:\n",
                   length(row.names)))
      base::show(objects(data))
    })
```

The method has no formal arguments. Fields `"data"` and `"row.names"` are referenced directly. As often happens in encapsulated methods, the method calls a function with the same name as a method of the class. To make the distinction, the function must be called with the package specified, here `methods::show()`.

Encapsulated methods inherit methods from all reference classes included in the `contains=` argument to `setRefClass()`, and from *their* reference superclasses. All reference classes inherit from class `"envRefClass"`. This class has a number of methods that act like default versions for standard computations.

The most important such method is `$initialize()`, which is called when a new object is created from the class, typically via a call to the generator function. Like the default initialization method for functional classes, it expects to get named arguments corresponding to the names of fields or unnamed arguments that are objects from a reference superclass. See Section 11.5, page 232 for examples of `$initialize()` methods.

A superclass method of the same name is called through `$callSuper()`. The arguments to `$callSuper()` are whatever the direct call to that method would require. The `$initialize()` method for `"dataTable"` has formal arguments (`...`, `data`) and calls its superclass method by

```
callSuper(..., data = as(data, "environment"))
```

There is no generic function involved and therefore no requirement that the methods have any particular argument list. Superclass methods only need to have the same name.

## 9.5 Example: Classes for Models

Two examples of classes to represent models show contrasting applications of functional and encapsulated OOP.

Models fitted to data are essentially functional: the object as a whole represents the functional result of some fitting algorithm. The fitted model object is not mutable: arbitrarily changing some property of it negates its definition.

On the other hand, models to simulate an evolving population or individual organism exist in order to change as a result of their (simulated) evolution.

Both model situations lend themselves to class definitions but the natural paradigm is a functional class in the first case and an encapsulated class in the second.

As an example of the first case, a linear least squares model will estimate coefficients and define residuals, optionally along with other information. A function creates the fitted model from arguments defining the form of the model and the data in question. Other functions can then display, analyze and extend the model.

The lm() function in the stats package returns an S3 class object of class "lm". This class and other models go back to the S version that introduced such classes [12]. The properties of the objects are not represented as slots or even, as it happens, attributes. Instead, the object is a named list with reserved names for the elements.

The S approach to models introduced "data.frame" objects for the data and "formula" objects for the structural model. From these, the fitting functions first constructed another "data.frame"—the "model frame" for the data actually used in fitting. Some numerical calculations would now fit a particular model of this class; in the case of lm(), a linear least-squares regression using a QR algorithm.

A modern approach to such fitted models could represent a similar structure formally. For example, here is a simple, minimal class for linear regression:

```
lsFit <- setClass("lsFit",
        slots =
          c(coefficients = "numeric",
            residuals = "numeric",
            formula = "formula",
            modelFrame = "data.frame"))
```

The four slots are those implied by the basic computational requirements, without assuming a particular numerical method (as the "qr" component of lm() does). Slot "modelFrame" is the model data frame, called "model" somewhat confusingly in [12]. That slot is implied by the use of formulas.

The QR numerical method used in the stats package would produce an object from a subclass of this, with slots to represent the numeric results. Alternative fitting computations would produce objects from alternative classes, but could continue to use methods (for example, for plot()) that did not depend on the numerical technique. Methods for functions that did depend on the numerical algorithm, such as for update(), would be defined for the subclass.

The value of generic functions and methods for users of software in this application is mostly through the uniformity of the function calls: calls to plot(), summary() and update() typically require little or no specialization for the user when applied to different classes of models.

The more technical advantages such as those that will be discussed in Chapter 10 have not found much application in this area. On the whole, most of the class/method structure for "old" types of models described in [12] has been modified in details but not fundamentally altered in later model-fitting studies, nor have features such as inheritance found really widespread use (an exception being the "aov" subclass of "lm" introduced in [12]). My guess is that the casual class definitions in [12] did not focus ideas on object-oriented extensions.

Let's turn to the encapsulated example. Here is a class definition for a simple simulation of an evolving population, which we used as an example of a naturally encapsulated OOP application (Section 2.5, page 38).

The model is just births and deaths in a population of identical individuals, reproducing asexually. For all individuals at each time point there is a single probability, the birth rate, that the individual splits into two and a single probability, the death rate, that the individual dies. The interest is plausibly to track the evolution of the simulated population, as determined by the birth and death rates and the initial population size. A later example (Section 16.2, page 342) has a slightly more realistic version in which different individuals have different birth and death rates.

The R definition of the class is:

```
SimplePop <- setRefClass("SimplePop",
                fields = list(sizes = "numeric",
                    birthRate = "numeric",
                    deathRate = "numeric"))
```

The population object is created with an initial size. Each call to its `evolve()` method simulates one generation of evolution and appends the resulting population size to the record of sizes in the object. Two methods suffice, the fundamental `evolve()` and a convenience method to return the current size.

```
SimplePop$methods(
        size = function() sizes[[length(sizes)]],

        evolve = function() {
            curSize <- size()
            if(curSize <= 0)
                stop("This population is DEAD")
            births <- rbinom(1, curSize, birthRate)
            deaths <- rbinom(1, curSize + births, deathRate)
            curSize <- curSize + births - deaths
            if(curSize <= 0) {
                curSize <- 0
                message(sprintf(
                "Sadly, the population is extinct after %d
                  generations", length(size)))
                }
            sizes <<- c(sizes, curSize)
        })
```

That's the whole model, with a few checks for extinction along the way.

# Chapter 10

# Functional Object-Oriented Programming

The plan for this rather long chapter is to describe in Sections 10.1 to 10.4 the use of functional OOP to extend R, covering most practical situations and including the essentials of the implementation. Sections 10.5 to 10.7 are references on details of classes, generic functions and methods. See the Index under "functional OOP" to find specific topics. Section 10.8 has some guidance on combining formal OOP with the earlier S3 implementation.

## 10.1  Functional OOP in Extending R

Functional OOP can contribute to extending R by extending the scope of the data we compute with or by extending the functional vocabulary. The first comes through defining a new class of data and methods for it, the second through a new function and methods to implement computations for it.

The main tools are functions `setClass()` new classes and `setMethod()` for new methods. The basic computations were discussed in Chapter 9. This section will look at what happens in the R process to make functional OOP work. The Sections 10.2 and 10.3 will deal with essentials of programming strategy for classes and for methods. For extending R, packages are once again the key (Section 10.4).

What "happens" in functional OOP is what happens in a function call, as always. The effect of the call depends on objects: the function object named in the call and the objects supplied as arguments.

R distinguishes generic functions as those for which methods have been defined. Following the OBJECT principle, the information used to evaluate a call to a generic function is found in the corresponding generic function object. This object

145

has a table of all the currently known methods for that function. Each entry in the table is a method object. Entries are indexed by the corresponding classes in the signature of the method definition.

Nearly all generic functions in R do nothing but *method dispatch*: selecting the method in the table that best matches the classes of the actual arguments and evaluating the body of that method in the frame of the call to the generic function.

The generic function object exists *in the* R *process*, and the table of methods reflects the dynamic nature of that process, as described in Chapter 5, in particular the ability to load R packages dynamically into the process. Methods for a particular generic function may be defined in several packages, and very frequently are (think of the methods for the `plot()` function). Class and method definitions in a package are brought into the R process by actions when the package is loaded.

If a package contains methods for a particular generic function, those methods will be added to the corresponding generic function object when the package is loaded into the R process. At any time, the method table in the function is determined by the methods in packages currently loaded into the process.

Both generic function objects and method objects inherit from ordinary functions, with some extra properties. The `"signature"` slot of the generic function object contains the names of the formal arguments for which methods can be defined. When a method is defined, it has a corresponding slot specifying classes for these arguments. The tables of methods in the generic function are indexed by method signatures. In practice, the tables and the selection process are simplified in several ways: Section 10.7, page 191 discusses the details.

The actual objects in a package containing class and method definitions are assigned with specially constructed names, to avoid conflicting with other assignments. The objects are created or modified during package installation (details in the following sections). The installation steps are as described in Section 7.2.

Class definition objects in a package are stored separately for each class. Method definition objects are stored within a single table for the corresponding generic function. The presence of method definition objects in the package's namespace triggers a load action to merge these into the generic functions in the R process.

To see the definitions, corresponding functional tools are provided: `getClass()` returns the class definition object; `showMethods()` lists the currently loaded methods for a generic function; and `selectMethod()` returns the method that would be dispatched for a specified function and argument classes.

One additional consequence of the R dynamic loading process is that methods can be defined in one package for a function in another even if that function has no methods in its own package. In this case, the original function may not be generic and often is not. For example, many methods exist for functions in the base package that are primitive functions there.

R handles this by the concept of an *implicit* generic function. The implicit generic version of a function has the same arguments as the ordinary function. By default, all arguments except for ... are in the signature. If that is not appropriate, an implicit generic can be specified (typically, to exclude some formal arguments but possibly to prohibit methods altogether): see ?implicitGeneric for details. The implicit generic mechanism is intended to work behind the scenes to provide consistency among methods in multiple packages. The default version is usually fine, unless some argument must not be evaluated in the standard way and therefore has to be excluded from the signature (think of the first argument of substitute(), for example).

Class and method definitions are specified in the R source for a package by:

```
setClass(Class, slots=, contains=, ....)

setMethod(f, signature, definition, ....)
```

as discussed in Sections 9.3 and 9.4. Calls to setClass() and setMethod() are not themselves purely "functional"; instead, their main contribution is to create objects in the package for the related class and method definitions. These objects are not visible or intended for direct use. They are *metadata* telling the evaluator what classes and methods are defined in the package.

The only arguments usually needed for setClass() and setMethod() are the first three shown above. The notation "...." stands for many additional arguments, few of which are of interest in ordinary use. setClass() and setMethod() date back to the initial S implementation of formal classes and methods. In the spirit of back compatibility, they still support the original programming style, some of which is not recommended in the modern R version. We will work through the recommended use starting with the most common situations.

Calls to setClass() will normally appear once per class to create new classes in the package. The rest of the programming mainly requires calls to setMethod().

One additional step sometimes required is to create the generic function itself or to declare that methods are about to be defined for an existing ordinary (non-generic) function in another package. The function setGeneric() is the main tool for this purpose (page 156).

## 10.2   Defining Classes

The class definition created by `setClass()` will specify the properties that objects from this class have and list any other classes that are its superclasses. In traditional object-oriented programming these two aspects are often referred to as specifying what an object *has* and what it *is*. The properties of functional classes in R are the slots, which are specified by name and by the required class for an object in that slot. The names of the slots and the corresponding classes for the slots are supplied by the `slots=` argument to `setClass()`. Superclasses are specified by the `contains=` argument, which is an (unnamed) list of other classes.

For example, there is a formal definition for the `"data.frame"` class, although that class is not usually treated formally. For a compact summary of the class, use the function `classSummary()` in the XRtools package:

```
> classSummary("data.frame")
Slots:
.Data = "list", names = "character",
    row.names = "data.frameRowLabels", .S3Class = "character"

Superclasses:  "list", "oldClass"
Package: "methods"
```

A data frame "has" a slot `"row.names"` required to be of class `"data.frameRowLabels"` (a class defined to allow the row labels to be numeric or character). It "is" a list, meaning that all list computations are valid for a data frame. The `".S3Class"` declares that this is a formal version of an informal, S3 class. The `".Data"` slot indicates that the data part of the object is a `"list"` vector. More details on these points later in this section and in Sections 10.8 and 10.5.

The class is created, in the methods package, by:

```
setClass("data.frame",
    slots = c(names = "character", row.names =
      "data.frameRowLabels"),
    contains = "list")
```

The class definition is an object, returned by the function `getClass()`:

```
> class(getClass("data.frame"))
[1] "classRepresentation"
attr(,"package")
[1] "methods"
```

The call to `setClass()` returns a generator function for the class. For example:

```
track <- setClass("track",
    slots = list(lat = "degree", long = "degree",
                 time = "DateTime"))
```

creates a class `"track"` and returns a function to generate objects from that class.

In addition to returning the generating function, the call to `setClass()` assigns the class definition as metadata. `setClass()` should be called directly from a file in the source of a package. The class definition is then assigned when the package is installed, with a constructed name. When the package is loaded, a load action will store the class definition from the metadata object in a table of known classes in the process.

A common programming style is to assign the generator function with the same name as the class, as above, which causes no conflict because the metadata object is assigned with a specially constructed name. Having a generating function with the class name is convenient for users and consistent with the practice in other OOP languages. However, it is only a convention and not required; for example, the `"data.frame"` definition did not follow it because there is already a `data.frame()` function and one would not want to hide it with the generator.

There is no requirement in fact to have a generator function at all, and earlier versions of the software had no special provision for them. In those versions, objects would be created directly by a call to `new()` supplying the class name. Generator functions are recommended because they are a little easier for users to type and also because they guarantee to specify the class unambiguously. While we have created a class `"track"` in this package, there is in principle no reason someone could not create a class of the same name in another package. If so, a user trying to work with both packages is guaranteed to get an object from the corresponding class by calling the correct generator function.

There are a number of other arguments to `setClass()` but only `Class`, `slots`, and `contains` are important for most applications. The others are mostly a combination of historical leftovers and special features that can also be specified by other tools. Just to get another historical leftover out of the way: `Class` is the first argument to `setClass()` but `slots=` and `contains=` need to be named explicitly, as we did in the outline on page 147.

## Slots

The slot names must be unique and the class definition for each slot must be available when `setClass()` is called. For another example:

```
binaryRep <- setClass("binaryRep",
    slots = c(
      original = "numeric",
      sign = "integer",
      exponent = "integer",
      bits = "raw"))
```

This specifies four slots with the names and classes given. The slots determine what data can be stored in objects of the class; an object, `b1` from class `"binaryRep"` must have a slot `b1@original` from class `"numeric"` or some subclass of that class. The slot may be assigned when the object was created by a call to the generator or by a later assignment with the `` `@` `` operator.

While every slot must have a class, that class can be specified as `"ANY"`, the universal virtual class, so in effect R has optional typing for slots in the usual sense of that term. Computations with the class are likely to be better defined by providing something more explicit, unless you really believe any R object makes sense in this slot.

Because the class for a slot must be defined before it can be specified in a call to `setClass()`, the source code for a package needs to define classes in the right order. For example, the class `"data.frameRowLabels"` in the definition of `"data.frame"` would have to precede that of `"data.frame"`.

The class for a slot may correspond to a basic data type, since these are built-in as classes. S3 classes are also valid, provided they are known to exist. For example, the formal class for linear regression models in Section 9.5:

```
lsFit <- setClass("lsFit",
         slots =
           c(coefficients = "numeric",
             residuals = "numeric",
             formula = "formula",
             modelFrame = "data.frame"))
```

Here, `"numeric"` corresponds to a basic type; `"formula"` and `"data.frame"` are S3 classes.

The one caveat for S3 classes is that they are undiscoverable on their own, since they have no definition or metadata object. To use them in functional OOP they just have to be registered as S3 classes. Registration is done by a simple call to `setOldClass()`. Many of the S3 classes occurring in the stats package have been registered, but some may have been overlooked, and S3 classes from other packages likely have not been registered. If one you want to use has not, your call to `setClass()` will fail with a complaint about an undefined class. The simple solution is to register the class in your own package. Class `"formula"` has been registered, but if it had not, a package wanting to use it just needs to include:

```
setOldClass("formula")
```

We'll say a little more about using S3 classes in Section 10.8, page 207.

One of the consequences of requiring slot classes to be previously defined is to exclude recursively having a slot that is itself from the class being defined. Such recursive structure is fundamentally at odds with the functional OOP paradigm, in which objects are not references. Suppose we tried to include a slot, say `"twinCity"` of class `"city"` in the definition of class `"city"`. This leads us to an infinite recursion, because the object in that slot must itself have a `"twinCity"` slot, and so on.

There are two approaches, one that cheats and a slightly more complex one that is safer and clearer. The cheating mechanism is to define the target class twice, in the first case as a virtual class. The second definition is the real one. R allows this (at least the current version of R does), only checking that the slot class is defined. One can avoid the infinite recursion by looking for a `"city"` object whose `"twinCity"` slot is NULL, because the prototype of a simple virtual class is NULL. The objection to this approach is that the NULL slot is in fact an invalid member of the class. Future versions of R might reasonably do a better job of validating the object and raise an error in this case.

For effective recursion in this form, the class union mechanism, discussed below, provides a similar but legal and understandable implementation of the same idea. As in the cheating version, we halt the recursion by making the slot be from some other class, indicating the end of the expansion. A `"NULL"` object is often used here as well, but now legally. The class specification is a union of class `"NULL"` and the original class.. It takes a little dexterity to make this happen; see the example on page 170.

Such recursive relationships tend to be inefficient if the code in effect iterates through many levels. They will be somewhat less so using encapsulated OOP, where the functional slot is replaced by a field in a reference object, less likely to require the cost of duplicating the object.

Specifying a slot as a class union or other virtual class is a useful technique more generally and widely applied, indicating that the object in the slot is required to be one of an explicit collection of valid classes (the class union case) or to have some behavior consistent with methods for the virtual class. The object in a slot specified as class `"vector"`, for example, can be from any of the basic vector data types or any subclass of these, while the virtual class `"structure"` implies both vector-ness and a structure that remains constant under element-by-element transformation [11, pages 154-157]. For more on virtual classes, see page 168.

## Inheritance

The slots of the superclasses named in the `contains=` argument also will be slots of the new class; that is, the slots are *inherited* by the new class, both the slot name and the required class for that slot. More than just the slots is inherited, however. Because an object from the new class *is* an object from the superclass, any method defined with the superclass in its signature is potentially inherited when an object from the new class appears as the corresponding argument. Where multiple methods are possible, the method selection procedure in R attempts to select the closest match (details in Section 10.7, page 197).

Inheritance of methods is probably the most powerful mechanism in functional OOP. It allows computational methods to be shared widely among classes of objects that are not required to be identical in structure, only to share the part of the structure critical to this computation There will be much to say along these lines as we look at strategies for effective functional object-oriented programming.

As with slots, the specified superclasses must be defined at the time of the call to `setClass()`. For example,

```
GPStrack <- setClass("GPStrack",
  slots = c(alt = "numeric"),
  contains = "track")
```

specifies class `"track"` as a superclass. If that class was defined as on page 149, its definition needs to precede this one in the package installation (page 114).

The classes specified in `contains=` are the simple, direct superclasses of the new class. "Simple" inheritance is an important concept in object-oriented programming in R. The assertion in saying, for example, that `"track"` is a simple superclass of `"GPStrack"` is that methods for the superclass can be applied to an object from the subclass, without additional tests or modifications. For classes defined by their representation, the assertion is trivial if the representation is inherited. In Section 10.5 we will examine class unions and other classes which are used for inheritance that does not follow directly from the representation. In these cases, simple inheritance relies on shared behavior that validates the inherited methods. The classic example is the `"vector"` class, which relies on all subclasses behaving like an indexable object.

In Section 6.1, page 97, some basic classes were discussed that correspond to the built-in data types in R. A class definition can specify one of these classes in the `contains=` argument; if so, then all the objects from this class will have the corresponding data type. So the `"data.frame"` objects will have type `"list"`, since that was specified as a superclass. In principle, any of the classes corresponding to a type can be specified. The objects containing method definitions, for example, belong to a class that inherits from `"function"` (type `"closure"`):

```
setClass("MethodDefinition",
  slots = c(target = "signature", defined = "signature",
    generic = "character"),
  contains = "function")
```

In addition to inheriting methods defined for the basic class, objects from the subclass will inherit behavior implemented in the core R code by examining the internal type of an object. For example, objects of class `"MethodDefinition"` can be called as a function, which is their main role in life.

No more than one such class can be a superclass of a given class. To express the idea that an object could be one of several types, define a class union (Section 10.5, page 169). If a class does not contain any of the basic types, objects from the class have the special type `"S4"`, to ensure that they are not treated as objects from a different, basic type. Objects from such classes are defined entirely by their slots.

Sometimes it is convenient to include the basic part of an object (usually referred to as the "data part") as if it too were a slot. This is the origin for that `".Data"` slot shown for the `"data.frame"` class. To get or set the basic object, without its class or slots, refer to the `".Data"` "slot".

```
> f <- selectMethod("show", "data.frame")
> class(f)
[1] "MethodDefinition"
attr(,"package")
[1] "methods"
> class(f@.Data)
[1] "function"
```

Objects with slots cannot have one of the reference types such as `"environment"`. Any attempt to assign to a slot in a local version would affect the original object. These types can still be superclasses, but the new class has a real slot for the data part. Programming can still treat them as having `".Data"`. Reference class objects are in fact environments. See page 177 for another example and details.

# 10.3 Defining Methods and Generic Functions

Method specifications are the most frequent part of programming in functional OOP, inevitably so given that a particular method corresponds to a combination of a generic function and one or more classes. A project is likely to require quite a few, whether the focus is on designing new classes or new functions.

The basic programing step is:

```
setMethod(f, signature, definition, ....)
```

This specifies that the function supplied as `definition` should be the method for the generic function `f` to be called when the actual arguments match the classes specified in `signature`.

The argument `f` is typically supplied as the quoted or unquoted name of the generic function. Either way, the function should be made available by being imported into this package. If the name is quoted, `setMethod()` will try to make it available, including promoting a non-generic version. It's cleaner though to be explicit by promoting a non-generic with a call to `setGeneric()`. If the function itself is defined in this package, the call to `setGeneric()` to create it needs to precede the `setMethod()` call in the installation. We'll look at `setGeneric()` on page 156.

The general rule that all objects from other packages should be explicitly imported is particularly important for generic functions. Two functions of the same name (from two different packages) may exist. The import makes it clear which is intended. It's possible that your package needs more than one of the functions (but very strongly not a happy situation, with the resulting confusion). In that unlikely case, the argument to `setMethod()` needs to indicate which function is meant, by including the package, in the form `PkgA::foo`.

The `signature` argument maps a specified class to each of a subset of the formal arguments to the generic function. Typically method signatures correspond to the first or first two arguments to the generic function, in which case the signature can usually be one or two character strings naming the corresponding classes. A general signature can be specified by any mixture of named and unnamed class specifications. All formal arguments not in the method signature are treated as if they were specified as class `"ANY"`. The rules for matching elements in the specified signature with the formal arguments are those used for matching arguments in R function calls. You can combine named and unnamed elements.

For any complicated situation, it's strongly recommended to generate the `setMethod()` arguments by calling `method.skeleton()` as shown on page 156. Even in simpler cases with multiple arguments, you should name all the elements of the signatures if there could be any ambiguity. Here is an example, taken from the rJava package interfacing to the Java language.

The generic function in this case is the operator `` `==` ``. This operator is a primitive function in the `base` package. These are treated specially. Most, including this one, can be treated as a generic function. A call to `getGeneric()` for a primitive function will return the generic version or signal an error if methods are prohibited.

```
> eq <- getGeneric("==")
> eq@group
[[1]]
[1] "Compare"

> eq@signature
[1] "e1" "e2"
```

The relevant information is that the generic function has two arguments in the signature, e1 and e2.

The rJava package wants to map the operator to the function .jequals() if either argument is from class "jobjRef", a reference to a Java object:

```
setMethod("==", c(e1="jobjRef",e2="jobjRef"),
          function(e1,e2) .jequals(e1,e2))
setMethod("==", c(e1="jobjRef"),
          function(e1,e2) .jequals(e1,e2))
setMethod("==", c(e2="jobjRef"),
          function(e1,e2) .jequals(e1,e2))
```

The second specification matches any call with a "jobjRef" first argument; the third specification matches any call with a "jobjRef" second argument. In the two cases, either argument e2 or e1 is omitted and so corresponds to "ANY".

Why include the first specification? All three are needed because a call with two "jobjRef" arguments is ambiguous otherwise. If only the e1 and the e2 methods were specified, which should be used if the call matches both? The evaluator would warn about the ambiguity. In this particular case, all three methods do the same computation but in general there is no guarantee that treating either argument as "ANY" gives the correct result. For details on such method-matching questions, see Section 10.7, page 197.

The argument name is commonly omitted for generic functions that only have one argument for which methods are specified. Again from rJava:

```
setMethod( "length", "jarrayRef", ._length_java_array )
setMethod( "length", "jrectRef", ._length_java_array )
```

This maps calls to length() to an internal function in the package, for either of the specified classes.

The definition argument to setMethod() should be a function with the same formal arguments as f. Method dispatch does not rematch the actual arguments to the arguments of the method. To do so would be inefficient and, more importantly, would render the method selection process incoherent. For convenience, some computational "sugar" allows extra formal arguments to be included in definition in addition to or instead of the ... argument. But this is only sugar: The effect is the same as defining the variant form as a function internally and passing the standard arguments to that function.

The initial source code for specifying a new method can be created reliably by a call to `method.skeleton()`. Even for fairly experienced programmers, this is a recommended way to create a call with the correct set of arguments and a valid method definition format.

For example, suppose we want to write a method for the `` `[` `` operator for `"dataTable"` objects (page 136), but only for character vectors selecting columns (because column order is meaningless):

```
> method.skeleton("[", c("dataTable", j="character"), stdout())
setMethod("[",
    signature(x = "dataTable", i = "ANY", j = "character"),
    function (x, i, j, ..., drop = TRUE)
    {
        stop("need a definition for the method here")
    }
)
Skeleton of method written to connection
```

The first argument is the generic function, the second the signature for which a method is to be defined. The third argument can be supplied as a file name for the output or omitted, in which case a default file name is used. It can also be a connection, as here or as an open connection to a file if you want multiple new methods defined in a single file.

Functional method objects created by calls to `setMethod()` are saved in metadata objects indexed by the generic function; so, for example, all the methods in a package for the function `show()` would be stored in a single metadata object, effectively a table of the methods. The methods in the table will be added to the corresponding generic function when the package is loaded into the R process.

## Generic functions

Functional methods are developed in packages but become active when installed in the function for which they are written. That function must be a generic function in R, meaning that it comes from class `"genericFunction"`, or a subclass. Actual method dispatch comes by calling the primitive function `standardGeneric()`. The definition of `show()`, for example, is:

```
function (object)
  standardGeneric("show")
```

This selects a method corresponding to the actual arguments, evaluates the body of the method in the frame of the current call and returns that value. Unless there is some exceptional circumstance, the call to `standardGeneric()` should be the *only* thing the generic function does. One of the exceptions is the function `initialize()` in the methods package:

```
function (.Object, ...)
{
    value <- standardGeneric("initialize")
    if (!identical(class(value), class(.Object))) {
        ....
    }
    value
}
```

Classes define methods for `initialize()` to specialize the creation of new objects when the generator for the class is called. The generic function checks that the object computed by the method has the same class as the supplied prototype object, `.Object`. Except for basic types, an error will be signalled otherwise. Such functions are nonstandard generic functions; the standard and nonstandard cases are distinguished by the corresponding class of the generic function object. If the function being called is a standard generic function the evaluator knows that method dispatch is all that's needed.

Functions such as `show()` and `initialize()` are imported from packages as generic functions, to which methods can be added by any package. In other cases, the package defining the methods may need to create a generic function, either corresponding to a non-generic function that we want to generalize, or as a new function that will have methods defined for it from the start. The technique in either case is to call the `setGeneric()` function.

None of the functions in the "older" parts of R, including the base, stats and graphics packages, is a generic function. A generic version compatible with the existing non-generic is created by calling `setGeneric()`:

```
setGeneric("plot")
```

The current definition becomes the default method and a generic version is assigned in the environment of the caller. For extending R, the call to `setGeneric()` will be in a package and the generic version will be assigned in the namespace of the package. Unless the existing non-generic function is in the base package, it should be imported by a directive in the current packages `"NAMESPACE"` file (even for common packages such as graphics.)

The generic is assigned in the application package but specified as *from* package graphics, according to its `"package"` slot:

```
> setGeneric("plot")
[1] "plot"
> plot@package
[1] "graphics"
```

Every package that creates methods for `plot()` in this way does so for the same generic function. When the different packages are loaded, the methods will all be added to the table of that generic function.

The step of creating a generic function by calling `setGeneric()` (explicitly or by calling `setMethod()`) seems a formality, but is in fact very important. It's worth understanding what happens and why.

The call to `setGeneric()` assigns a generic function object in the namespace of the package at install time. With this package loaded, there will be at least two objects named `"plot"` present in the R process, and potentially more, if multiple packages have methods for `plot()`.

Quite naturally, some users extending R find this situation unnerving. But in fact it is the key to the functional, object-based approach to OOP in R. The essential points are that the generic function is consistent with the function in the original package and that all packages following this procedure will create *identical* generic function objects. Methods can be written for the function in multiple R packages, all consistent with the original function and dispatched from one corresponding generic function in the R process. At the same time, all packages importing the original non-generic function will continue to call that object from any functions in their namespace.

Regardless of the number of packages using either version, there will be only two distinct `plot()` objects in the R process and the generic version will dispatch methods from one table, containing all the methods defined in packages currently loaded into the process.

Some of the non-generic functions have been written to use the older informal classes and methods, typically referred to as S3 classes and methods. The "non-generic" function in this case will in fact be selecting and dispatching S3 methods: its definition will typically be a call to the function `UseMethod()`.

The `plot()` function is an example; its definition in the graphics package uses S3 methods:

```
plot <- function (x, y, ...)
    UseMethod("plot")
```

The SoDA package for the examples in [11] defines some formal methods for `plot()`. A `setGeneric()` call in that package assigns the generic version. *Any* package defining methods for `plot()` will have the identical generic function.

```
> setGeneric("plot")
[1] "plot"
> identical(plot, SoDA::plot)
[1] TRUE
```

Packages that import the **graphics** version will load and use the non-generic function. These packages may define S3 methods for `plot()` to be called by the S3 mechanism for methods.

Packages that, like **SoDA**, use formal methods will load and use the generic function. Again, they may or may not define methods of their own. For example, a package with no methods of its own might import the generic function from package **SoDA**. Packages that themselves define methods may have an (identical) function in their own namespace.

Understanding the roles of the generic and non-generic functions can clarify situations that might be confusing. Consider the `as.data.frame()` function in **base**. This function uses S3 methods, with a slight variant, to convert an object to a data frame. The widely used function `data.frame()`, also in **base**, applies `as.data.frame()` to each of its arguments and combines the results into a single data frame. Formal methods for `as.data.frame()` are an attractive idea. A call to `setGeneric()` will create the generic as usual.

But the `data.frame()` function will continue to call the non-generic version, as it should. We might have hoped that defining methods for `as.data.frame()` would cause those to be used in calls to `data.frame()`, but that will not be the case as is clear from our analysis of the two versions of the function. We will look at this example in Section 10.8, page 207.

The solution shown there, to define an S3 method, works and is defensible if the old function, `as.data.frame()` in this case, dispatches S3 methods. In fact selecting methods, either in functional OOP or in the older S3 implementation, is the one place where functions modify their computations dynamically during the R session. If `as.data.frame()` did some fixed computation, not based on a dynamically changeable set of methods, then functional programming says its definition should indeed be constant, not subject to change from another package.

Once again, we are at the intersection of functional programming and OOP here. Keeping the two in harmony given the dynamic nature of method tables in R requires a form of "moral restraint".

Going back to the `plot()` example, packages such as **SoDA** should only define methods for classes that the package "owns". This is indeed the case for **SoDA**, which has methods for two of its own classes.

Implementing a method in this package for class `"lm"`, say, would be a violation of functional programming, for either a formal or an S3 method. The `"lm"` class is defined (by implication, since it's an S3 class) in package **stats** and there is a corresponding S3 method for `plot()` in that package. Defining an overriding method would potentially alter the behavior of a computation in another package even though that computation made no use of the SoDA package. We will formulate this moral requirement generally on page 162.

All the above concerns the case that the function exists. The `setGeneric()` call announces that we plan to define methods for it. Sometimes an application may want to create a new function and define methods for that. For example, the XR package for interfaces described in Chapter 13 defines a function `asServerObject()` whose purpose is to replace an arbitrary R object with some text in another language that evaluates to the equivalent object. This is something new, but clearly asks for methods depending on the R object and perhaps on the particular server language as well.

The function is created by a call to `setGeneric()` that includes a function definition as a second argument:

```
setGeneric("asServerObject",
    function(object, prototype) {
        ....
    })
```

The second argument is the function definition to use; its formal arguments and the expressions for the default values, if any, become those of the generic function. The function supplied becomes the default method for the generic function. In the source for a package, this computation needs to come before any `setMethod()` calls for this function. For the details of creating new generic functions see page 187.

For much more detail in general on generic functions, see Section 10.6, and on methods, see Section 10.7.

# 10.4    Classes and Methods in an R Package

We are assuming throughout this chapter that the software being designed will be part of an R package; indeed, the amount of thought typically required when designing new classes and methods would be wasted if the resulting software was not easy to reuse and to share. Long before reaching this level of programming-in-the-large, a package is a necessity to make good use of our efforts.

Compared to a package consisting only of function definitions and/or objects with data, a package including classes and methods requires a few extra steps.

- Classes, methods and generic functions are all created when the package is installed. The source code for the package must be set up so those computations work. In particular, R needs the superclasses of a new class to be defined before the class itself, and warns if methods are defined before the classes specified in the method signatures.

- If the `"NAMESPACE"` file for the package tries to be explicit about what is exported, rather than using a standard pattern, then one may need some explicit export directives for methods and classes. (Similar import directives may be needed for other packages that depend on this package.)

Installation of R code for a package begins by concatenating the source files into a single file for parse-and-eval. By default, the files are collated alphabetically by file name.

The requirement for classes being defined in order is then most easily satisfied by putting all the class definitions into a single file (with the class definitions in the file in a valid order), and then choosing the file name so it collates before other files in which methods are defined including the defined class in their signature.

A common programming style in encapsulated OOP puts a class definition along with all the methods associated with that class into a separate file. You can do that, but in this case you must arrange that the files are "collated" in the right order. The collation order can be made explicit in a `Collate:` directive in the `"DESCRIPTION"` file. Unfortunately, this requires that *all* source files be named in the directive; any file not mentioned will not be included in the installation.

## Loading with classes and methods

In step 9 of loading a package as described in Section 7.3, the namespace of the package is examined for class and method metadata objects. These are specially named objects assigned in the package as a side effect of calls to functions such as `setGeneric()`, `setMethod()` and `setClass()`.

There will be one such object for each class defined in this package and one for each generic function for which the package has any methods. The metadata information will be used to update two tables in the R process, one for class information and one for method information. The two tables are indexed by references to classes and to generic functions, respectively. The class table will be used to validate slots or superclasses when computing with an object from the class. In the table of generic functions, each function has an internal table which will be used to select a method when evaluating a call to that generic function.

These are tables of references which as always in R means that each entry is identified by a name and an environment, in this case implied by the name of a package.

All of this information is part of the R process. It's key to understanding how the evaluator works. In particular, notice that all the information is potentially dynamic and capable of changing as namespaces are loaded or, less often, unloaded.

The definition of a generic function essentially *is* its table of methods. For any function, this can change when a namespace is loaded having a method for that function. A package can have methods for any generic function that it imports. These methods will be merged into the table of generic functions at load time.

In addition, calls to the function cause methods to be identified using inheritance when necessary. For example, if the function had a method corresponding to class `"vector"` but none explicitly for `"numeric"`, a call with a numeric argument would find the `"vector"` method. Once found, the method is cached in the table to eliminate the need for future searches. We'll look at the details of method selection in Section 10.7, page 197.

The class definition can be dynamic also, not in its actual content but in its relation to other classes. A class will get new subclasses if the package has classes that inherit from this class. It can also get a new *superclass* as well, but only in the special case that a class union in the new package has this class as one of its members.

### The right to write methods

In encapsulated OOP, methods for objects from a class unambiguously belong to the class (or to a superclass). Functional OOP has a richer but more complicated relation among generic functions and class definitions.

A generic function is defined by the set of all methods. But at the same time, the behavior of a class takes in all the methods in which it appears. With general method signatures, this includes behavior that depends potentially on the function and on more than one class.

A `plot(x,y)` method for plotting an object `y` from one class against an object `x` from another is part of the definition of each class. At the same time, if the method is sensible, it is consistent with the intent of the generic function; if it does not contribute to the implementation of the function's intended purpose, it should not be a method for this function.

The generic function or the classes in the signature may come from more than one package. In this case, the "ownership" of a potential method for this combination is shared.

As a hypothetical but not ridiculous example, suppose we are dealing with data from the `"GPSTrack"` class defined on page 149 and at the same time with some matrix classes representing similar data, using classes of data defined in the Matrix package. It would be convenient to define some methods for mixed operations on the two classes. The ownership of these methods is in a real sense shared, since their definition uses information about classes in both our package and Matrix.

The ownership may be shared, but it is still meaningful and it's important for reliable computing that it be respected. A method in one package for which the generic function and all the classes in the method signature are owned by other packages is a "rogue" method. It can corrupt computations that have nothing to do with that package.

As noted above, loading a package into the R process adds the methods from the package to the current version of the corresponding generic functions. Rogue methods can change the result of a computation as a result of loading the package it belongs to, even though the computation has nothing to do with that package.

Consider our hypothetical example. It's possible that some of the methods in the Matrix package might not match our use of those computations in our application, where the matrices have a different interpretation. The temptation is to replace the methods by more appropriate versions. This would be a very bad idea. Our package owns none of the classes, nor the generic function. Defining any such method could change the behavior of computations in other packages simply as a result of loading our package. For computations producing important numerical information, the result could be undetected and potentially disastrous errors.

The moral is clear: Methods should only be written in a package when the package is home to one of the class definitions or to the generic function.

# 10.5 Functional Classes in Detail

This section explores a number of questions of strategy:

- decisions about class inheritance;

- using virtual classes (page 168);

- classes that extend the basic R data types (page 172);

- specializing object initialization (page 178);

- functional classes for reference objects (page 181).

### Class inheritance; the "data.frame" example

*To inherit or not to inherit?* That is a question that arises often in defining new classes. Inheritance is one of the most powerful concepts in OOP. When one would like to add some functionality to a class, or some additional information (new slots) to the objects, defining a subclass of the class is usually the best idea. In most cases, defining a subclass is a better idea than trying to add to the existing one. With R packages, the original class may be in another package, meaning that you cannot modify its definition in that package (and should not on grounds of programming clarity in any case).

It's not necessary that the new class have additional structure. It may have the same information as the original but imply some constraints; for example, `"matrix"` is a subclass of `"array"` that constrains the length of the `"dim"` slot to 2.

Inheritance can be used also to define a "mixin", a class whose objects combine the features of two, usually unrelated classes. R allows mixins, just by specifying more than one class in the `contains=` argument to `setClass()`. Numerous examples exist, but there are some downsides, particularly if the two superclasses are not entirely unrelated. Method selection, for example, may find multiple valid methods by following different paths through the inherited class structure (for details on method selection, see Section 10.7, page 197).

To examine the consequences of inheritance, one must examine the various methods including the superclass in their signature. Do these make sense for the new subclass? If they do not, how does one define a method that overrides the inherited one? Does the existence of undesirable inherited methods suggest conceptual issues with the inheritance itself?

Let's look at the `"data.frame"` class as an example and compare it to an alternative without inheritance, looking in some detail at how this very popular class has been integrated with R.

The `"data.frame"` class as implemented in R is an S3 class, but is equivalent in structure to a formal class definition:

```
setClass("data.frame",
    slots = c(names = "character",
            row.names = "data.frameRowLabels"),
    contains = "list")
```

This is a vector class, since its data part is a list vector. With no special methods the built-in methods for list vectors would be inherited. For many classes that extend vectors, these inherited methods are appropriate in at least a substantial subset of applications. For data frames, more methods need to be overridden, including those for the `` `[` `` and `` `[<-` `` operators used to extract or replace subsets.

The data frame inherits from `"list"`, but also extends the matrix concept with the addition that columns can be of different classes. While the *implementations* of data frames and matrices in R have little in common, their functional behavior (the expressions one would use for them) is frequently similar. One often thinks of row- and/or column-subsets of either, for example. At the same time, some matrix operations such as transpose make little sense for a data frame.

As in many applications, the choice is between broad inheritance, which produces many computations automatically with some of them a bad idea, and narrow inheritance (or none at all), which produces few computations automatically so that the new class only gradually becomes useful. The right choice will depend on how close the superclass' behavior is to the new class and on what computations are likely to be applied. One is also choosing here between a relatively "safe" design for the new class versus greater convenience provided by inherited methods, at the risk of incorrect results.

The original, informal `"data.frame"` class is likely the most widely used non-elementary class of objects in R. This makes it a useful example for illustrating the good and bad effects of inheritance.

The standard software for R has over 50 informal methods for the `"data.frame"` class:

```
> methods(class="data.frame")
 [1] $                $<-            Math            Ops
 [5] Summary          [              [<-             [[
 [9] [[<-             aggregate      anyDuplicated as.data.frame
[13] as.list          as.matrix      by              cbind
[17] coerce           dim            dimnames        dimnames<-
[21] droplevels       duplicated     edit            format
[25] formula          head           initialize      is.na
[29] merge            na.exclude     na.omit         plot
[33] print            prompt         rbind           row.names
[37] row.names<-      rowsum         show            slotsFromS3
[41] split            split<-        stack           str
[45] subset           summary        t               tail
[49] transform        unique         unstack         within
see '?methods' for accessing help and source code
```

Some of the methods are substantial: the method for the `` `[` `` operator is nearly 150 lines long, that for `` `[<-` `` nearly 300. The programming effort involved would be hard to justify if the class were not so central to many computations. We can ask what the goals were, and perhaps end up with helpful analogies for introducing other class definitions. Looking over the list of methods, most satisfy one or more of three criteria:

1. to supply information you would likely want for any useful class (plotting, printing, summarizing, constructing objects);

2. to provide functionality for the class analogous to a similar class from which this class does not inherit, in this case `"matrix"`;

3. to avoid calling a method for a superclass of this class, when that method is not considered to be suitable (in this case, methods for `"vector"` or `"list"`).

The first category comprises five or so of the methods for the class. The large majority satisfy one of the other criteria, in roughly equal numbers, and quite a few satisfy both. In the end, relatively few default methods are inherited for functions likely to be used with data frames.

The `"data.frame"` class would inherit the vector methods for `` `[` `` and `` `[[` ``. These are overridden by `"data.frame"` methods for the operators, but largely to allow matrix-like versions. In effect, the informal methods attempt to make data frames inherit from matrices, for these operators and for other functions. In this sense, the class is an example of the mixin mechanism—a class that inherits from two separate classes. But in fact, `"list"` and `"matrix"` are not unrelated classes; both inherit from `"vector"`. The warning flag mentioned above for mixins is up: If the two inherited classes are related, difficulty in defining methods can be expected where they have overlapping behavior. The complexity of the methods for `"data.frame"` operators and the subtle distinctions arising that often confuse users are hints of just such problems.

Consider the `` `[` `` operator for extraction, for example, as in the expression:

```
iris[1:3]
```

If the data frame `iris` were interpreted as a matrix, this would return the first three elements of the first column, by the column-wise ordering of elements. If interpreted as a list, the expression returns the first three variables (or "columns"). Existing methods use the list interpretation in this case, whereas with

```
iris[1:3,]
```

they follow the matrix interpretation, indexing the first three elements of each variable. While the motivation for the current choices is understandable and the current code is very unlikely to change, confusion for users is likely.

For programmers writing methods, the consequences are even more subtle. In both expressions above, argument `j` to the operator is missing. Therefore, another mechanism must be used to distinguish the number of arguments. In the first example this is two, in the second three (with the third argument empty).

The existing informal methods for these operators interpret variable or column indexes with no restrictions. In particular, numeric indexing of variables in the data frame goes ahead just as it would for a list. But does this make sense for model fitting?

Nothing in the conceptual paradigm for data frames says that the numerical index of a variable has a special meaning. Erroneous software might result from methods that use variable ordering. For example, I may have a convention that the first variable in the data frame is the dependent variable. But if I want to write software that assumes this, it should not be given out to be used on arbitrary data frame objects. (Instead, I should define a subclass of data frames that includes this assumption and write methods using that class.) From this perspective, a better version of the `[` and `[[` operators might allow only indexing on the variable names.

A different argument leads to the need for methods for the replacement versions of the operators, `[<-` in particular. The inherited method from class `"list"` is in this case the default internal computation. A problem arises because this method will allow users to modify the object so as to make it invalid as a data frame. The internal method, for example, makes no distinction with respect to the length of the replacement object, but assigning a variable of the wrong length must be avoided for data frames. An explicit method must be defined to override the inherited one. Indeed there is one for the informal `"data.frame"` class, nearly 300 lines long.

```
> length(deparse(getS3method("[<-", "data.frame")))
[1] 286
```

The majority of the code sorts out varied argument patterns, interpreting them according to one or the other of the two implicitly inherited superclasses and detecting prohibited cases.

It is helpful to compare the `"data.frame"` class with a variant having no automatic inheritance. We can do so with minimal changes to the actual representation:

```
setClass("dataFrameNonVector",
  slots = c(
    data = "namedList",
    row.names = "data.frameRowLabels"
  )
)
```

The essential change has been to move the list to an ordinary slot from the special `".Data"` slot representing the data part of the object. The `"names"` slot has moved with it, but we have done this implicitly by giving class `"namedList"` to the `"data"` slot.

If the non-inheriting class definition were used for the same model-fitting computations, a different design decision is needed. If no method is provided for iterating over variables, the software that needs to do such iteration is required to know more about the definition of class `"dataFrameNonVector"`. Iterating over variables is arguably a pretty basic operation on such objects, and it's generally good design to hide the details of such basic operations. One reason is that the class design may then modify how the computation is done without invalidating the existing software that uses the operation. An R-like way to provide this particular operation would be a variable-wise `apply()`-type function. The `"data.frame"` class inherits the internal version of `lapply()` in the `base` package. As a result, `lapply()` can be used for `"data.frame"` objects.

An `lapply()` method for `"dataFrameNonVector"` would be a natural way to provide iteration over variables. The definition can be as simple as:

```
setMethod("lapply", "dataFrameNonVector",
          function(X, FUN, ...)
              base::lapply(X$data, FUN, ...))
```

Computations with the class can then also use `sapply()`, since it is written in terms of `lapply()`.

Where the `"data.frame"` class had to override most of the inherited subset operator methods, class `"dataFrameNonVector"` does not allow any subset operations by default. If these are important, a method is needed and this method faces many of the same design issues considered for `"data.frame"`. If model fitting was the main goal, one might take the attitude that the operators should have been applied to the data from which the current object was generated, for example a data table from a reference class such as the one in Section 11.3. Users who do want to manipulate the data then must learn about other classes but the software for this class may be cleaner and more reliable.

The central role of data frames in statistical computing gives us an especially good chance to compare strategies, but similar questions should be asked about whatever class design is likely to be important in a project to extend R. The important step is to imagine as well as possible the likely uses of the software, in the large, and provide a good combination of flexibility and trustworthiness.

## Virtual classes; class unions

Following the OBJECT and FUNCTION principles, we can think of a class in R in terms of what an object from that class *is* and what happens to it, what it *does*. What the object is comes from its class definition, in the form of required slots. What it does comes from what functions can have the object as an argument; that is, from the methods for which objects from the class are eligible.

For a generic function, the desired computations might be expressed in terms of some common behavior. If the object could implement that behavior, a method for the function could be defined. Writing a method whose signature includes a class representing the shared behavior is then an effective programming mechanism, but such a class is only partially defined. No object will actually have this as its class.

Functional OOP in R provides a formal mechanism for such partially defined classes. Such classes are *virtual*. A virtual class is incomplete and therefore no objects can be created from it. Methods can be defined with a virtual class in the signature, however. Other classes that inherit from the virtual class can use those methods. Programming for all the subclasses is then made easier.

The original example in R is that of vector objects. The term "vector" refers to objects for which indexing by numeric or logical expressions is meaningful to extract or replace elements. The concept that certain objects share this "vector" behavior has always been part of R (and S before). Basic types such as `"double"`, `"character"` and `"list"` are specific vector classes. But knowing that an object is a vector is not enough to define the contents of the object; one must know the data type as well for the various basic vector classes. No object has `"vector"` as its actual class.

The most common and usually the recommended way to define a virtual class is as a *class union*. Class unions are entirely defined by listing their direct subclasses; there is no content requirement implied by inheriting from the class union. They are quite special classes, because any existing class can be included in the union, making the union a superclass of the existing class. Because there is no implication about content, the implications are only for methods having the class union in their signature. These methods assert that they work for all classes that are members of the union. For the informal class union `"vector"`, the assertion is essentially that the built-in methods for subset and element operators work, along with some related functions such as `length()`.

The declaration of the class union gives its name and then all the initial members of the union. A definition of `"vector"` as a class union would look like:

```
setClassUnion("vector",
    c("logical", "numeric", "character", "complex",
      "integer", "raw", "expression", "list"))
```

(Having been around for much longer than class unions themselves, the actual vector class in R is not defined as a class union.)

In the formal definition of the `"data.frame"` class (on page 164) a class union is used for row labels:

```
setClassUnion("data.frameRowLabels",
    c("character", "integer"))
```

This just expresses the policy that either integers (typically starting as `1:nrow(x)`) or character strings can be used as the vector of row labels. Methods could be defined for this class if the computations worked for either of the member classes.

A useful limiting case for class unions arises if we want to designate some special object in an ordinary class; for example, an object that implied "the object you're looking for hasn't been defined". The natural way to indicate an undefined state in R is as the null object, `NULL`. If one wants to write methods that apply to a class but check for the case that the object has not been defined, a union of that class with class `"NULL"` is convenient. The class `"OptionalFunction"` provides for a slot that may have a function but may also be "empty":

```
setClassUnion("OptionalFunction",
    c("function", "NULL"))
```

Methods for the class union will arrange to create an object from an actual class in the union by testing the argument with `is.null()`.

If objects are passed as references, one could define some special object as the empty value instead of allowing `NULL`, but that won't work with ordinary R objects, including functions. These are not reference objects, so it's only by the contents that one could detect such a special case. This always risks constructing the "empty" object accidentally. In addition, testing with `is.null(x)` is both simpler to program and clearer when reading the code than a test for a special `"function"` object would be. For reference classes in R, using a special object does work and is used in the XR package.

New classes can be added to an existing class union by specifying the union in the `contains=` field in the call to `setClass()` If the new class is a subclass of one of the member classes of the union, it does not need to be added to the union in order to use the existing methods. While the basic data types are the direct members of `"vector"`, specialized subclasses of one of the basic types will still use vector methods, unless one chooses to override these.

One specialized case where extending a class union can be useful is to provide for the kind of almost-recursive slot definition we mentioned on page 151.

Suppose we want a slot in the class `"city"` that is either another `"city"` object or `"NULL"`; in other words, a recursive version of the optional object case. One approach is this:.

```
setClassUnion("optionalCity", "NULL")
setClass("city",
    slots = list(location = "site", name = "character",
        twinCity = "optionalCity"))
setIs("city", "optionalCity")
```

The key is to define the union first, then the new class, then add the new class to the union, all in order to ensure that each class has a definition when it's used.

Outside of such special cases, adding a class to a union can be dangerous. If the union was really just a convenience for dealing with a fixed set of possible classes, adding a different class to the union requires guessing all the common characteristics used. Software using the class union may in fact just be switching between the members of the union, in which case adding another class is likely to either be ignored or generate an error.

The virtual class `"vector"` is an example. This was intended to group together all the basic data types. Adding a new vector class with, say, a different mechanism for storing the data would require methods for a very large and poorly defined collection of functions. Such a modification would be challenging.

Where expanding the class union is problematic, a slightly more complicated approach is safer: Define a new class union from the old union and the proposed new members:

```
setClassUnion("data.frameRowLabelsExtended",
  c("data.frameRowLabels", "myLabels"))
```

No changes are required for methods involving the original class union.

Class unions have one special privilege. The member classes may include classes whose definition is sealed. Class definitions are sealed to prevent other packages or user code from redefining them and potentially invalidating methods or other code using the class. In particular, all the classes for the R data types are sealed. Including one of these classes as a member of a class union technically violates the seal. It adds the union class to the direct superclasses. So, for example, the `"data.frameRowLabels"` class union has changed the direct superclasses of `"integer"`, as we can see from the class definition.

```
> whichSuperclasses("integer", 1)
[1] "numeric"                "vector"
[3] "data.frameRowLabels"
```

The definition of `"data.frameRowLabels"` has added that class to the superclasses of `"integer"`. (See package XRtools for the function `whichSuperclasses()`.)

Without class unions functional OOP in R would be less useful, and class unions would be much less useful without the privilege of including existing classes as members. But bad things can result:

```
> ## No one would do this of course, but ...
> setMethod("summary", "data.frameRowLabels",
+           function (object, ...)
```

```
+                    "Nothing")
[1] "summary"
> summary(1:10)
[1] "Nothing"
```

The method definition has modified the behavior of a function in the base package on a basic data type, something that one should never do. Future versions of R may be able to detect such egregious misbehavior; even now, the bad act would have had no effect if the function had been a primitive (sum(), for example) instead of a true function such as summary(). Primitives only look for methods if their arguments come from a formally defined class.

If such bad stuff was in the package where "data.frameRowLabels" was defined, blame would fall on that package and presumably no one would use the offending class. Otherwise, the fundamental sin in this example is disobeying the "right to write methods" principle (page 162). We did not own either the class union or the function summary() and therefore had no right to write a method for that combination.

Occasionally, one wants a virtual class that not only prescribes some behavior but also implies some particular structure, such as one or more slots that a subclass must always have. In this case, an arbitrary class definition can be made into a virtual class by including "VIRTUAL" as one of the classes that this class contains. Suppose we wanted all the classes that share some time-stamp behavior to have a slot "timestamp" with class "DateTime". We can express this by having all such classes extend a corresponding class, say:

```
setClass("TimeStamped", slots = list(timestamp = "DateTime"),
        contains = "VIRTUAL")
```

All classes defined to contain "TimeStamped" would have the required slot.

## Basic R data types

Data types are built into the internals of R. They map directly into the C implementation, encoded as a field in the C structure common to all R objects. Core computations in R contain some form of switch over the possible data types. Historically, they go back to the original versions of the language, in the original implementation of S (where they were called modes).

Objects can be generated from classes corresponding to basic data types, but the objects do not actually "have" the relevant class stored in the object. The class is defined by implication from the data type.

For most purposes, such as method dispatch for regular R functions, the distinction is not important. Note, however, that the R utility functions is.object()

and `isS4()` that detect an object with a class or an object from a formal class will both return FALSE for such objects. (These functions are detecting a bit set in the C structure, and correspond to quick internal tests for dispatching methods on primitives.)

```
> l1 <- new("list", list(1:10, rnorm(10)))
> class(l1)
[1] "list"
> is.object(l1); isS4(l1)
[1] FALSE
[1] FALSE
> attributes(l1)
NULL
```

Primitive functions do not check for formal methods if none of their arguments has the corresponding bit set.

Having used the `"list"` data type and class in this example, we should note a detail for these objects that can be confusing. It is traditional in R to think that the list data type optionally comes with a `"names"` attribute, which can be a character vector of names for the elements of the list but may also be missing. Although this convention is traditional and used by non-OOP code, it complicates software in that the attribute is optional, and so needs to be checked repeatedly. Although the `names()` function can return NULL, it's not the case that the attribute is optionally a NULL object, but that there is no such attribute.

In treating data types as formal classes, R is consistent in the rule that the basic vector classes have no attributes. Class `"list"` has no names attribute; lists with names have their own class, `"namedList"`, for use in class and method definition. This is a formal class, not a basic data type:

```
> n1 <- new("namedList", list(i=1:5, x=rnorm(5)))
> class(n1)
[1] "namedList"
attr(,"package")
[1] "methods"
> is.object(n1); isS4(n1)
[1] TRUE
[1] TRUE
> n1
An object of class  "namedList"
$i
[1] 1 2 3 4 5
```

```
$x
[1] -0.9432269  0.2811897  0.7781412 -0.3972630 -0.4683104
```

## Classes that extend R data types

Although the data types don't have an explicit class attribute, they can be used like other classes, either as the class for a slot or as a class in the signature of a method. One of these classes can also be used as the superclass of another class. In this case, the superclass is referred to as the *data part* of the subclass, the sense being that the new class acts like data of the corresponding type, but with additional properties. This is historically a very old idea in S and R, predating any formal use of classes.

The data type determines the internal representation of the object. For example, the various vector types imply that the data consists of a C array of some corresponding C type. Other types have other internal representations. None of this is meant for the programmer to use explicitly, but having a specific type as the data part of a class causes objects from the class to share one of these representations, and therefore inherit software written for that data type.

Methods are inherited, including core computations that are not formally methods for the particular data type, but which adapt internally to different types, often making for efficient computations for arithmetic, comparisons and much else. If the class definition does not extend such types, objects from the class will have their own data type instead, "S4". As one might expect, not much in the way of built-in computation will happen for objects with type "S4", other than manipulating slots or dispatching programmer-defined methods. The tradeoffs are similar to our general discussion of inheritance: the convenience of inherited computations balanced against the danger that some of those computations may be undesirable or wrong.

Objects with data of an existing R type can come from a class definition that extends one of those types, such as:

```
setClass("datedText", contains = "character",
         slots = list(date = "DateTime"))
```

The class may also inherit from a class union whose subclasses include data types. A built-in example is the class "vector", with all the vector types as subclasses.

```
datedVector <- setClass("datedVector", contains = "vector",
         slots = list(date = "DateTime"))
```

In the second example, the type of an object from the class is not fixed, and will be whatever type corresponds to the actual object stored as the data part. The possibilities are those shown on page 169.

Classes can inherit the data type implications through multiple generations: a class that inherited from `"datedText"` would also have type `"character"`. The class definition in methods assembles all the implied superclasses to establish the type.

A class definition can allow a choice of types but a specific object can only have one; an attempt to specify more than one generates a warning and uses one of them:

```
> setClass("myDataWithTrack",
+      contains = c("numeric", "character"),
+      slots = list(coords = "track"))
Warning message:
In .validDataPartClass(clDef, where, dataPartClass) :
    more than one possible class for the data part: ....
```

If what you meant was that either of the two choices was allowed, define or use a suitable class union (page 168):

```
> setClassUnion("numberOrText", c("numeric", "character"))
> setClass("myDataWithTrack",
+      contains = "numberOrText",
+      slots = list(coords = "track"))
```

Using a union for the data part does require that methods for the class work correctly for all members of the union.

We have used the term "data part" somewhat loosely to refer to the class that determines which data type will be used. For formal classes, the convention is to treat this as a special slot with a reserved name.

The convention is that slot `".Data"` corresponds to the data part as determined by the inheritance. If we look at the class defined by the previous example:

```
> classSummary(getClass("myDataWithTrack"))
Slots:
        .Data              coords
Class "numberOrText" "track"

Superclasses:   "numberOrText"
```

The explicitly specified superclass defining the data part appears also in the `".Data"` slot. If the data part was inherited indirectly, the two entries in the summary will reflect that:

```
> summary(getClass("dataWithGPS"))
Slots: .Data = "vector", coords = "GPSTrack"
Superclasses:  "structure"
```

Actually, the `"structure"` class is built in, but it behaves as if it were a class with `"vector"` as its data part.

## Extending reference types

So far, the data types used as the data part have been "normal" R data types, in that objects are treated functionally. Objects of the various vector types can be passed into function calls and modified locally there without having side effects on the original object. Not all R objects have this behavior. Some types of objects behave according to a reference paradigm, so that changes are reflected everywhere.

Classes exist corresponding to reference data types, as they do for normal data types. These include `"environment"`, `"name"` (corresponding to the data type `"symbol"`), `"externalptr"`, `"builtin"` and `"special"`. Of these, `"environment"` is by far the most relevant. The data type is the basis for reference classes, is fundamental to R evaluation and is useful on its own as a mechanism for defining tables. We discussed environments generally in Section 6.3. As with normal data types, objects generated from these classes will still be basic R objects, with no attributes and not flagged as S4 objects:

```
> abcName <- new("name", as.name("abc"))
> is.object(abcName); isS4(abcName)
[1] FALSE
[1] FALSE
> attributes(abcName)
NULL
```

As with normal data types, these classes are legitimate either as slots in class definitions or in the signature of method definitions.

Reference types can also be used as a data part, but an indirect implementation is required. R does not allow the object "reference" to appear with different attributes. Changing an attribute such as the class is reflected everywhere, making direct extension of reference types disastrous. (See the discussion in Section 6.2, page 104.) In particular, S3 classes cannot extend reference types and the S4 classes only appear to have an actual reference type. An internal mechanism uses a different reserved slot, a real slot this time. The internal code dealing with these objects operates to treat the object in that slot as if it were a data part.

We can extend the examples with date slots to data of class `"environment"`:

```
> setClass("datedEnvironment", contains = "environment",
+          slots = list(date = "DateTime"))
> classSummary("datedEnvironment")
Slots:  date = "DateTime", .xData = "environment"

Superclasses:  ".environment"
```

There are two differences in the class summary from the previous examples. Instead of a ".Data" slot we have ".xData". While ".Data" is a conceptual slot standing for the data type of the object, ".xData" is an actual slot. Secondly, whereas the direct superclass was the same as the class for the ".Data" slot, now it is not; for example, ".environment" rather than "environment". Both these differences result from an internal mechanism that keeps the reference object in a separate slot, so as not to corrupt it, and defines inheritance to extract that slot. For your programming you don't need to use the internal mechanism (and you should not). Just treat the data slot, ".Data", as you would with other types, but don't use the type directly (not usually a good idea in any case).

Quite a few basic computations that treat objects from the formal subclass will automatically pick off the slot, at least for classes with "environment" in the data part. You can't count on it, so be sure to check and do a coerce, as(x, "environment") in this case, when in doubt.

A more substantial application would be to develop a data-frame-like class that used an environment rather than the named list in the "data.frame" class. As an alternative strategy for providing the functionality of data frames, the use of a reference object has considerable appeal, if also some dangers. By being a reference object, the environment implementation will not automatically duplicate the variable as would be the case for a regular data frame. Recall the "data.frame" class definition:

```
> getClass("data.frame")
Class "data.frame" [package "methods"]

Slots:

Name:                    .Data               names
Class:                    list           character

Name:              row.names             .S3Class
Class: data.frameRowLabels           character
```

```
Extends:
Class "list", from data part
Class "oldClass", directly
Class "vector", by class "list", distance 2
```

One simple analogue using an environment would be:

```
dataEnv <- setClass("dataEnv",
    contains = "environment",
    slots = c(row.names = "data.frameRowLabels"),
)
```

The `"row.names"` slot has stayed the same, the `"list"` in the data part is now an `"environment"` and the `"names"` slot has disappeared, since the names of objects in an environment are not stored in an attribute, but kept internally.

## Initializing objects

All objects from functional classes are generated by a call to `new()`, which creates a standard object from the class and then calls `initialize(.Object, ...)` with that object as the first argument, followed by optional arguments passed to `new()`. Methods written for `initialize()` allow the class designer to specialize the new objects to the needs of the class. The default method for `initialize()` expects named arguments for the individual slots or unnamed arguments containing objects from a direct superclass.

Initialize methods can be useful tools in class design but they are also somewhat tricky to implement. In this section we will discuss the general strategy, with some pointers to avoiding common defects. The design principals apply largely to the `$initialize()` methods for encapsulated OOP as well. The `$initialize()` methods differ mainly in assigning fields by name rather than replacing slots of `.Object`. For details of the encapsulated case, see Section 11.5, page 232.

Before looking at some examples, some general comments and cautions need to be stated. Motivations for writing an initialize method may include: recasting the arguments describing objects in a more convenient form than the explicit slots or fields; validating relationships or other properties required for constructing the object; modifying or extending similar initialization features inherited from one or more superclasses.

Some design requirements for initialize methods need to be kept in mind. We'll state a few first, then illustrate with an example.

- Since methods for `initialize()` will be inherited by subclasses of the class currently being defined, the method should allow for slots not part of the current definition. In practice, this can usually be handled by calling the next method (superclass method for reference classes) after processing any arguments treated specially in this method.

- In principle all non-virtual classes should be capable of returning a default object from the class when `new()` is called with no optional arguments, and in the current implementation such default calls can be generated when sub-classes are defined. For this reason all special arguments in `initialize()` methods should handle the case of a missing argument in some way.

First, a simple example. Suppose class `"m"` has a slot `"data"` that is declared of class `"matrix"`. The default initialize method would require specifying `data=` that was a matrix; providing a vector as the slot would cause an error. An `initialize()` method to relax the requirement by coercing an argument explicitly to `"matrix"` would be:

```
setMethod("initialize", "m",
    function(.Object, data, ...) {
        if(!missing(data))
            .Object@data <- as(data, "matrix")
        callNextMethod(.Object, ...)
})
```

The method shown tests `missing(data)` rather than having an explicit default value, which leaves the class definition free to specify the prototype for this slot in whatever way makes sense. Notice that the initialize method now allows the `data` argument to be supplied unnamed, as in:

```
m1 <- new("m", 1:10)
```

One might prefer to require the `data` argument to be named, rather than matching it positionally. That preserves the usual rule that only superclass objects are allowed as unnamed arguments; to implement this, the only change needed is to put `data` after ... in the argument list.

Validity methods are defined by a function of one argument, an object from the class. The call returns either `TRUE` if the object seems valid or else some text describing the problems found. A validity method can be defined in the call to `setClass()` or specified separately for readability by calling `setValidity()` later in the package source. A validity method will be called automatically from the default method for `initialize()`. The recommended form of an initialize method ends with a `callNextMethod()` call, to ensure that subclass slots can be specified in a call to the generator for the class. If this convention is followed, initialization will end with a call to the default method, and the validity method will be called after all initialization has occurred.

The actual call to the validity method comes from a call to the function `validObject()`. Except for the call from the default `intialize()` method, validity methods are not called automatically in standard R computations, partly for efficiency reasons but also because some modifications are likely to require objects that are temporarily invalid. For example, if two slots are required to be of the same length, then updating first one and then the other would fail if each slot assignment triggered a validity check. Because the checks are not done automatically, it's a good idea for methods and functions that update an object with a specialized validity requirement to finish with a call to `validObject()` as a check.

In the example on page 149 defining the class `"track"`, the slots for latitude and longitude were specified as class `"degree"`. The intent of this class is to guarantee that the object, structurally just a numeric vector, has numerical values for the co-ordinates in a specified range. This is formalized in a validity method:

```
degree <- setClass("degree",
                   contains = "numeric")


setValidity("degree",
    function(object) {
        nbad <-
            length(object) -
                sum(is.na(object) |
                    (object >= -180. & object <= 180.))
        if(nbad)
            gettextf("%d values out of range +/- 180.",
                     as.integer(nbad))
        else
            TRUE
    } )
```

Organizing the requirement in terms of a validity method simplifies imposing the constraint in other classes. The `"track"` class can incorporate the constraint in its own initialize method by calling the `degree()` generator function:

```
setMethod("initialize", "track",
    function(.Object, lat, long, ...) {
        .Object@lat <- degree(lat)
        .Object@long <- degree(long)
        callNextMethod(.Object, ...)
    })
```

Simply declaring the class of the slots would have guaranteed the same result, but in that case users of the "track" class would have been required to call degree(). The implementation shown maintains the validity requirement, but without extra work for users.

## Reference objects and functional classes

While all data types and other classes can be used in nearly all contexts, using objects with a reference paradigm inside a functional OOP class can raise some issues that you should consider. Inside a reference class, we expect the fields to behave as references; that is, changing a field in an object changes the object wherever it is used, not just in the function call responsible for making the change. But in a functional OOP class, changing a slot that behaves like a reference changes that slot and *only* that slot, wherever the object is used. Problems arise if changes to the reference part of the object can be made in one function call but are not anticipated in a concurrent function call. In a simple case, suppose: the functions f1() and f2() both use an object x that has some reference behavior; f1() calls f2(); and f2() changes the reference part of x. Then f1() is in danger of making erroneous assumptions about x. Let's take this as the description of the problem; the same issues will arise if there are other function calls between the f1() and f2() and of course there may be more functions playing the roles of these two in the computation.

Reference data types are the most obvious candidates for such problems, and our examples in this section deal with them. However, other objects in R have reference behavior, but without explicitly using that term. Connections, for example, will read from or write to a common stream such as a file. If the connection object is shared between f1() and f2() and f2() uses the connection, f1() needs to take this into account. Just this sort of bug has come up many times in practice.

Problems only arise if f2() modifies the reference information and f1() can not or does not take account of the modification. For some applications, one can localize all the modifications inside one function, f1() in our terminology. If all other functions treat x as read-only, then computations should be consistent.

If that is not the case, some extra care needs to be taken. A reasonable strategy is to use fully reference-style information, specifically by defining a reference class for the application. Then f1() has access to all the information about the current state of x, and just needs to be careful to use it. If you do not want to use a reference class, it is possible to simulate the necessary behavior by using replacement functions. In other words, all calls to f2()-style functions must be replacement assignments to x, in the style of f2(x) <- value.

Let's illustrate with an example. Consider the "datedEnvironment" class on page 177. It has the reference class `"environment"` as its data part and the ordinary class `"DateTime"` as the `"date"` slot. Suppose the idea was to record the last date of modification in the `"date"` slot. A utility function to update a particular entry in the table might then be:

```
setEntry <- function(object, what, value) {
    object[[what]] <- value
    object@date <- Sys.time()
    object
}
```

This uses the method for ``[[<-`` inherited from `"environment"` to store `value` under the name given by `what`, and then sets the `"date"` slot to the current time. But the `"date"` slot is changed only inside this function. Unless the calling function replaces its version of the entire object with the value returned by `setEntry()`, the new date will be lost, but the changed entry corresponding to this update will still be there. And any other function up the expression stack will also see the revised entry but the old date. Is this what the programmer intended? It seems unlikely.

The danger here comes from mixing functional and reference behavior. They do mix if you are very careful, but more often you will want to choose one or the other. In this example, if the table was really viewed as a reference object, which the use of class `"environment"` suggests, then a natural implementation would be as a reference class. Something like this, perhaps:

```
datedTable <- setRefClass("datedTable",
    fields = list(table = "environment", date ="DateTime"),
    methods = list(
      entry = function(what, value) {
          table[[what]] <<- value
          date <<- Sys.time()
      })
  )
```

In the reference version both the table and the date must be fields, but reference behavior applies to both. Changes to the table in any function will have the form

```
tbl$entry("X", newX)
```

Our function `f1()` does need to be sure that it uses the current contents of the `"date"` field (by referring to `tbl$date` directly, for example).

If one wanted to use the `"datedEnvironment"` class as it was defined, all calls to functions that might modify the object must be replacement operations.

We wrote `setEntry()` in the form of a replacement function: it takes the intended right side of the replacement as its final `value=` argument and returns the modified object. All that's needed is to call it `` `entry<-` `` instead of `setEntry` and to ensure that all the calls from `f1()` have the form

```
entry(tbl, what) <- value
```

For simple situations, this solution works fine. If the object can be modified in different ways by different functions or the modification may be done several layers of function call down, then the replacement approach becomes complicated and error prone.

The whole computation could also be cast in a functional form, perhaps less obvious for this particular example, but consistent in the changes made by the `setEntry()` utility. All that actually needs to be altered in the class definition is to use a functional class instead of `"environment"`, the natural choice being `"namedList"`. A functional form would be:

```
setClass("datedList", contains = "namedList",
        slots = list(date = "DateTime"))
```

Now all the changes will be made, consistently, to the local object, but if one wanted all the objects to have the same information, then the replacement approach must still be used, with its potential difficulties.

# 10.6  Generic Functions in Detail

A generic function in R can be viewed in terms of what it does, when called, or in terms of what it is, as an object. In the first sense, a function is generic if a call to it results in a call to a method for that function, with the same actual arguments as the call to the generic function. The method will be selected from a table of all methods currently registered for this function, the selection being that method best matching the classes of the objects supplied in the call. In the second sense, it is an object from a subclass of the virtual class `"genericFunction"` and also inherits from the class `"function"`.

Generic function objects come from one of three classes (or from one of their subclasses).

`"standardGenericFunction"`: a function that does nothing but select and call a method, with the value of the method call returned as the value of the function call;

`"nonstandardGenericFunction"`: a function that calls a method but does some other calculations as well;

`"groupGenericFunction"`: a function that is not called directly but which has a set of group member generic functions, with the methods for the group generic eligible to be selected for a call to any of the member functions.

All these classes are subclasses of the virtual class `"genericFunction"`.

## Generic functions during an **R** session

Generic function objects have a `"signature"` slot, containing the names of the formal arguments that can be included in the signature of a method. Packages with `setMethod()` calls for this generic function will contain a table of corresponding methods. During an R session all the currently active methods for the function are stored in the function, in a table indexed by corresponding classes for the arguments in the signature. This organization of methods is the key to understanding what a generic function does, so we need to examine it in some detail.

All generic functions are associated with an R package. As with all R objects, the reference to a generic function is the combination of a name and an environment (here the package, via its namespace). Generic functions are slightly different in that R arranges to merge methods intended for the same generic function when those methods come from different packages. For this purpose, both the name of the function and the name of the package are part of the object itself (in the current implementation, literally so via slots `"generic"` and `"package"` in the object). From now on, when we talk about a generic function in an R session, we will always mean one identified by the combination of function and package names.

Allowing several packages to define methods for the same generic function is an essential part of functional OOP in R. Software from many packages provides the greatest part of the value of R. Packages that define classes and methods often provide methods for generic functions from other packages. This is most obvious when the methods are for "built in" functions to manipulate objects, do arithmetic or logical operations, and other basic computations. But R allows methods in a package to be defined for a generic function from an arbitrary other package. In an R session, there will be only one generic function combining a given generic name and package name, and containing all the corresponding methods currently loaded into the session.

The distinction between a generic function during a session and the original function definition in a package is important to understand. The R process has a table of all active generic functions; in turn, each of these functions has a table of its currently known methods. When a package is loaded into the session, all the methods defined in that package are inserted in the corresponding method tables. *All* calls to that generic function will use the same set of methods, regardless of where the call comes from. Keep in mind that a "generic function" here corresponds to the combination of function name and package name.

A few typical examples may clarify the discussion. Methods for the function `initialize()` from the **methods** package define how new objects from a particular class are initialized. Class definitions arise naturally in many packages. The single version of `initialize()` in an R session will have all the methods defined for such classes. Notice that classes also are associated with a package, so that in principle two classes with the same name from different packages are free to each have its own `initialize()` method.

The subset operator `` `[` `` is another common candidate for methods. In this case, the operator is built in as a primitive, but still accepts methods. Methods for this operator are usually defined corresponding to the class of the object, `x`, but may in addition depend on the class of the index, `i`, or even on other arguments. All such methods will, again, be stored in one version of this generic function in the R session.

A third common function for methods is `plot()`, the base graphics function to generate `x-y` plots. This function is defined as a non-generic in the **graphics** package. Methods are frequently defined for a single argument, corresponding to `signature(x = "myClass", y = "missing")`, but can be defined for any combination of suitable argument classes.

The three examples of generic functions illustrate three different conditions: in the first case a function that is only defined as a generic, in the second an R primitive and in the third a non-generic R function treated as an implicit generic function when methods are defined. The three cases are treated differently in some details but behave consistently in selecting methods.

In terms of our distinction between encapsulated and functional OOP, notice that methods in this model are clearly not "encapsulated" in the definition of a single class. There is no context-free definition of "all the methods" associated with a particular generic function. It's legal, natural and encouraged that programmers develop packages to extend the definition of a generic function. Whenever a user loads that package, the corresponding methods are added to the current definition of the generic.

The actual dispatch of a method for a function, say `f()`, is executed by the call:

```
standardGeneric("f")
```

This selects a method according to the arguments to the calling function, executes a call to the selected method with the same arguments, and returns the value of the call to the method. The argument to `standardGeneric()` is the name of the generic to use for dispatch. The argument is really a historical leftover from S, since it must be the same as the name of the generic function in all normal usage. Here, "executes a call" is a slightly special computation: Arguments are not rematched to the formal arguments of the method. Instead, a new calling environment is created with the same arguments and the body of the method is evaluated in this environment. The parent environment of the call is that of the method; specifically, the namespace of the package in which the method is defined.

The method selection process allows methods to be inherited. The details of that selection are discussed in Section 10.7, page 197. Once the inherited method has been determined, it is cached in the generic function's table so that subsequent calls with the same signature do not repeat the inheritance computation.

### Implicit generic functions

The generic version of an initially non-generic function is created automatically either by a simple `setGeneric()` call or as a side-effect of a `setMethod()` call. The generic function is the *implicit generic function* corresponding to the ordinary function.

The default definition for the implicit generic function is the previous non-generic version, so that calls with arguments that do not match any explicitly specified methods will behave as they did before.

The implicit generic function is entirely defined by the non-generic function and the package in which it was found, an essential requirement for distributing methods over multiple packages. The generic function created has a package slot corresponding to the original function regardless of what package contained the `setGeneric()` call.

For example, the Matrix package defines methods for a function `kronecker()` which has a non-generic version in package base. This creates a generic version in Matrix:

```
> find("kronecker")
[1] "package:Matrix" "package:base"
> class(base::kronecker)
[1] "function"
> Matrix::kronecker
standardGeneric for "kronecker" defined from package "base"

function (X, Y, FUN = "*", make.dimnames = FALSE, ...)
standardGeneric("kronecker")
    . . . . . . .
```

Suppose another package defined some methods for `kronecker()`. If this package and **Matrix** are both loaded, calls to `kronecker()` must select methods consistently with all the available methods. They will do so because the two packages have the identical generic function, namely, the implicit generic function for function `"kronecker"` and package `"base"`.

In the example above and in the majority of cases, the implicit generic function is inferred automatically from the original, non-generic function: The generic function has the same arguments as the original and methods can be defined for all formal arguments except `"..."`. Just occasionally, the package chooses to specify a different rule for the implicit generic. To do so, the original package stores the implicit generic definition in a special table. The paradigm for this in the source code for the package is to create the generic function, in the usual way, and then to set the function back to the non-generic version by calling `setGenericImplicit()`. For example, on page 191 we will explain the need for an implicit generic function corresponding to `with()` in the base package and show the mechanism involved. The required definition is stored in the table, and the non-generic version restored, by:

```
with <- function (data, expr, ...)
    eval(substitute(expr), data, enclos = parent.frame())
setGeneric("with", signature = "data")
setGenericImplicit("with")
```

Notice that a new specification of an implicit generic must be in the package containing the non-generic version of the function, *not* in one of the packages wishing to define methods for the function, in order to ensure that all such packages deal with a consistent version of the function. (An exception is that the **methods** package is responsible for implicit generic functions in the **base** package and the other built-in packages that were written without acknowledging OOP software.)

## New or explicit generic functions

Not all generic functions start from the desire to generalize an existing ordinary function. A generic function can be created for a new purpose in one step, by supplying the default version as a second argument to `setGeneric()`. Package **XR**, discussed in Chapter 13, defines two new generics, `asServerObject()` and `asRObject()`. We showed the first on page 160; the call to create the second is:

```
setGeneric("asRObject",
            function(object, evaluator) {
                object
            })
```

This is a function to convert a simple object returned from an evaluator in a server language into a potentially special form.

The second argument to `setGeneric()` is used to construct a standard generic function with the same argument list. The function itself becomes the default method, which in this case just returns the argument `object`.

One might want a function that is defined for certain combinations of classes but has no default definition. In this case, the second argument to `setGeneric()` should have a body that just calls `standardGeneric()`. For example, package Matrix has function `expand()` which expands certain condensed storage forms. It makes no sense to call the function except for objects from one of the corresponding classes; `expand(x)` makes no sense if `x` is an ordinary vector, for example. Package Matrix defines the generic for `expand()` as:

```
setGeneric("expand",
   function(x, ...) standardGeneric("expand")
   )
```

This generic function will have no default method, unless one is explicitly created.

```
> showMethods("expand")
Function: expand (package Matrix)
x="CHMfactor"
x="MatrixFactorization"
x="denseLU"
x="sparseLU"
```

Methods exist only for some of the classes in this package; calling `expand()` with other objects will generate an error. If there is a reason for not having default methods, it would be helpful instead to define one that explains the problem, say:

```
setMethod("expand", "ANY",
  function(x, ...)
    stop(gettextf(
      "Expand only makes sense for factorizations, not for %s",
      dQuote(class(x)))))
  )
```

It is possible, but not usually a good idea, to create explicit generic functions even if there is a non-generic version; that is, generic functions that are different from the non-generic in arguments or their default method. If we wanted a generic version but decided that the existing function was not suitable as a default or if the generic should have a different argument list than the non-generic, an explicit generic can be used. However, if all that was wanted was to exclude some arguments from the possible method signatures, that does *not* require a distinct generic; the signature can be specified explicitly or included via the implicit generic mechanism, as we will see on page 190.

Creating a generic function by a two-argument call to `setGeneric()` always associates the generic with the package containing the call. If it's a new generic this is natural. But if the call redefines a function in another package, there is potential for serious confusion. Generally, you should avoid doing this.

As an example (of discouraged behavior), suppose one decided for some reason to eliminate the **decreasing** argument from the function `sort()` in base.

```
> ## THIS IS NOT A GOOD IDEA, BUT:
> setGeneric("sort", function(x, ...) standardGeneric("sort"))
Creating a new generic function for 'sort' in package 'myPackage'
[1] "sort"
```

The point to notice is that the generic is not from **base** as it would be with a simple call, but from the new package. It is a *different* function, which is the main reason not to do this.

Method specifications for function `sort()` now must be clear which generic is intended. There is considerable potential for errors, depending on whether the non-standard version has been imported to another package or is being found on the search list from **base**.

Better in nearly all situations is to create a differently named function that is clearly an alternative, such as:

```
setGeneric("mySort",
           function(x, ...)
               base::sort(x, decreasing = FALSE, ...)
           )
```

It's also worth noting that the calls to `setGeneric()` with two arguments do the same thing as assigning the non-generic version and then calling `setGeneric()` with one argument. A one-expression definition is a little clearer and probably the better choice.

Methods for functions that have ... in the argument list can have named arguments in addition or instead of ..... The mechanism is to define a local function inside the method with the modified argument list and to call that function from a constructed method definition; see [11, pp 393-396] for details. As a slight variation on the `mySort()` generic, for example, here's a version that removes the **decreasing** argument from the generic but retains it for the default method (so default calls that worked with `sort()` still work with `mySort()`):

```
setGeneric("mySort",
           function(x, ...)
               setGeneric("mySort")
           )
## but leave the default method alone
setMethod("mySort", signature(),
          function(x, default = FALSE, ...)
              base::sort(x, default, ...)
          )
```

This has the effect of making `mySort()` behave identically to `sort()` whenever the new function does not have an applicable method.

## Signatures for generic functions

A generic function object has a `"signature"` slot, specifying which of the formal arguments in the function can be used in method selection. By default, the signature consists of all the formal arguments with the exception of `"..."`, if the function includes that argument. For example:

```
> formalArgs(kronecker)
[1] "X"              "Y"              "FUN"
[4] "make.dimnames" "..."
> kronecker@signature
[1] "X"              "Y"              "FUN"
[4] "make.dimnames"
```

The "dots" argument, `"..."`, is excluded because its behavior is highly non-standard, essentially substituted for the name `"..."` when that appears in calls from this function Unlike other formal arguments, it is not an object in the environment of the evaluation of the call. There is a special mechanism for methods using this argument (see the online documentation `?dotsMethod`), but it does not mix with other arguments.

The signature can be set explicitly, nearly always because one of the arguments to the function is not processed according to standard R behavior; and that in turn is usually when the argument is not evaluated, but used symbolically. I would not recommend introducing more examples of such arguments—they cause a variety of headaches and limit the use of code tools—but a number of such functions do exist, for example the function

```
with(data, expr, ...)
```

This function evaluates the expression in the second argument, in the environment or other context defined by the first argument. However, it does not evaluate `expr`, but treats it as a literal:

```
> with(list(a=1, b=2), a+b)
[1] 3
```

Ordinarily, `a` and `b` would need to be defined where the call to `with()` originated.

The distinction matters for functional methods because the requirement is that all the arguments in the signature of the generic function be evaluated. It would not be possible to define methods for `with()` depending on the class of `expr`. So the signature of the generic version of `with()` contains only `"data"`.

As it happens, `with()` is one of the functions in the base package, and these are not explicitly generic functions. So the implicit generic function mechanism is used. The methods package in R has in effect the code:[1]

```
with <- function (data, expr, ...)
   eval(substitute(expr), data, enclos = parent.frame())
setGeneric("with", signature = "data")
setGenericImplicit("with")
```

Follow this three step pattern whenever a special implicit generic definition is needed: 1-define the non-generic version; 2-define the generic with its signature or other special requirements; 3-store the implicit generic and restore the non-generic function.

# 10.7  Functional Methods in Detail

The key to understanding functional methods is that they are just what their name says: Methods for computing the value of a call to the corresponding generic function. A generic function essentially searches for the best match to the relevant classes of the arguments among the available methods and then evaluates the body of that method.

As discussed in Section 10.6, every generic function in an R session has a table of the currently known methods. A method is dispatched by looking in the table for an entry that corresponds to the signature of the current call; that is, to the classes corresponding to all the relevant arguments in the call. If there is no such entry, the evaluator searches for an eligible inherited method, which it uses and installs in the table, so that the inheritance search is not needed on a later similar call.

---

[1] We're cheating a little; the actual definition includes an S3 generic. See Section 10.8.

If there is more than one eligible method, the evaluator looks for one that is preferred to all others. If no unique "best" method exists, the evaluator notifies the user of all the possibilities, but selects one method (essentially arbitrarily). If there is no eligible method, an error results.

The selected method is evaluated in an environment initialized with the same arguments as for the call to the generic function. Arguments are not re-matched but otherwise the computations then proceed as they would have for a direct call to the method as a function. In particular, the environment of the method as a function is the namespace of the package in which the method was defined, not the namespace of the package for the generic function. Methods can use any object reference in the package's namespace and imports.

For a standard generic function (and most generic functions are standard) selection and evaluation of the method is all that happens. A call to the generic is a call to the method. Some generic functions are nonstandard, and do additional computations before and/or after evaluating the method. In these functions the call to `standardGeneric()` does the standard method evaluation, returning the value of the method. For example, the function `initialize()` is a nonstandard generic. It examines the result returned by the method to check that it is actually an object from the target class.

Method selection—the choice of a method from among those that could validly be used—is the most important detail of functional methods. The design of classes and the methods themselves, taken together, should ensure that appropriate methods are selected.

## Method selection: examples

Our next goal in this section is to understand how the design of classes and methods determines method selection. R does have a rule that will pick an eligible method for each function call, perhaps with a warning about ambiguities. We will look at that rule in some detail, but first at two examples that illustrate how a collection of methods for some function needs to work together.

Nearly always the key design goal is to embed what *should* happen clearly in the class and method formulation, to avoid ambiguous or non-intuitive method selection for users (not to mention erroneous results). Specifically, each likely combination of arguments in the function call should map to a specific best-matching method. That careful design, and not exploiting the more arcane aspects of method selection, is the essential take-away.

Functional OOP is fundamentally more complicated than encapsulated OOP. Methods belong to functions, but are indexed within each function by classes of (some of) the arguments to the function. This requires a table indexed by keys each of which contains several values, when more than one argument is involved. That in itself is not a big deal, and if the objects in a call to the function had to match the specified classes exactly, life would be simple.

However, inheritance is one of the essential power tools for OOP. We want a method to be available to objects from subclasses of the specified class as well. The specified class can inherit from more than one superclass. And when multiple arguments to the function can appear in the signature of the method, inheritance is relevant for each of these. Therefore, more than one valid method may exist. A method is valid in general if, for every argument in the signature, the method was specified for the corresponding class of the object in that argument or for a superclass of that class.

In some cases, one valid method is clearly preferable to another. For example, if one method was defined for the actual class and another for a superclass, then the first is preferred, other things being equal. R's method selection follows a particular decision rule to determine which method is considered best, or if multiple choices qualify. This section will describe that rule.

In fact, although it's usually swept under the rug, inheritance raises non-trivial issues for method selection in encapsulated OOP, to a lesser extent. We'll note this below when we look at method selection for a single argument, which is analogous to what encapsulated OOP does.

We will use two contrasting examples where method selection is important. The first is the simple and common requirement of methods for binary operators, which we looked at previously on page 155 for the `==` operator in package rJava.

Binary operators for built-in R data types and standard extensions such as matrices are usually primitives in the **base** package. Many extensions to R naturally want to implement these with other classes of data, both on their own and mixed with existing R classes. The rJava example wanted to compare Java objects with each other and when relevant with regular R objects.

Our other recurring example comes from the XR structure for interfaces discussed in Chapter 13. Here we have two new generic functions, each with two arguments. One argument is a potentially arbitrary R object and the other represents either an object in a server language or the evaluator for that language interface. The goal here is to provide standard methods along with the ability to customize those for new languages and for particular classes of objects. We looked at the generic function definitions on pages 160 and 188.

The binary operators are special in that they correspond to generic function *groups* [11, 403-405]. All the arithmetic operators belong to group generic function `Arith()`, all the comparison operators to `Compare()` and these two functions in turn belong to group generic `Ops()`.

The **methods** package defines a formal version of the vector `"structure"` class

that has been around informally since S began. This allows the class to be used in
a consistent way in functional OOP. As part of the definition, there are methods
for binary operators implementing the concept of vector structures.

Most operators work element-by-element and the structure concept is that
element-wise operations leave the structure intact if the other operand is a simple
vector. (Think adding a scalar or a vector of the same length to a matrix.) This
applies to all operators, arithmetic or comparison, and is reflected in methods for
the `Ops()` generic. With a structure on the left and a vector on the right:

```
setMethod("Ops", c("structure", "vector"),
          function (e1, e2)
          {
              value <- callGeneric(e1@.Data, e2)
              if (length(value) == length(e1)) {
                  e1@.Data <- value
                  e1
              }
              else value
          })
```

This first computes the result for the vector part via `callGeneric()`, which in
this case will call the primitive for the particular operator. If the result has the
same length as the structure, the assumption is that it has operated element-wise
and a corresponding structure is returned. Otherwise (e.g., if the vector was longer
than the structure), only the vector part is returned.

There is an identical logic for the case of a vector on the left and a structure
on the right. Finally, a method is defined for two structures. In this case, one could
examine the two structures for being identical, but that would be time-consuming
and not entirely reliable. Instead, the method simply throws away all structure
information:

```
setMethod("Ops", c("structure", "structure"),
          function (e1, e2)
          {
              callGeneric(e1@.Data, e2@.Data)
          })
```

The assumption is that specialized classes extending `"structure"` may implement
specialized methods for this case.

The pattern for most implementations of operator methods will be similar:
three methods, one for each mix of the new class with general objects and one for
two objects from the new class. This is required even if the computations are not

distinct in all three cases. Method selection can only judge by what exists, not by an analysis of what the individual methods do. So a missing method for the two-object case leaves selection ambiguous, as noted in the rJava example. There all three methods called the same function but needed to be set explicitly. (The rJava case is somewhat non-functional in design in that the comparison operator always returns a single value, unlike most uses of this generic.)

The "structure" example is on the simple end; for contrast, consider the Matrix package, one of the recommended packages with R. The package implements classes for a variety of matrix structure (sparse, dense, triangular or symmetric) and corresponding vectors, all crossed with data types for the elements where relevant. This requires a prodigious number of methods, as we can see using a utility in the XRtools package. The eligibleMethods() function returns a character vector identifying all available methods for particular generic functions, optionally restricted by signature or package. This gives a count of the methods in the Matrix package for Ops() and Arith()

```
> length(eligibleMethods("Ops", package = "Matrix"))
[1] 71
> length(eligibleMethods("Arith", package = "Matrix",
+                   doGroups = FALSE))
[1] 66
```

Even with 130+ methods not all combinations can be unambiguously determined. We'll look at one case as an example of the method selection procedure.

The binary operators are central parts of R. The challenge for a new set of methods is to adapt those to a range of object classes in a complete form that resolves all combinations.

The XR example introduces a new pair of generic functions along with a standard implementation. The generic functions are intentionally open-ended in that they have an arbitrary R object as one argument. The packages that extend XR for particular languages may find the standard implementation adequate, may augment it for various special cases or may replace much of it with different methods. The challenge is to end up with a consistent set of methods. Ambiguous choices are only an issue for defaults and usually easy to fill in. More of a challenge is to obtain the right results in all likely applications.

The computation supported by the two generic functions is to evaluate an expression in the server language (Python or Julia in the two example packages) and interpret the result as an R object. The function asServerObject() translates an R object into a string that, when parsed and evaluated, is asserted to produce an equivalent object in the server language. The server language evaluator is expected to translate the result of the evaluation into a string that, when interpreted in R, creates an R object. These objects are only defined for a fixed range (basically scalars and lists, optionally named lists). The asRObject() function turns these objects into the full range of possible R classes. The XR strategy includes a way of coding arbitrary R objects; the decoded version is then used in a second call to asRObject(). That technique allows methods to deal with R features not easily captured in the general representation.

The XR package implements all this using the JSON notation (details in Chapter 13), but without any specialization for the server language. Five methods for asServerObject() and four for asRObject() are defined:

```
> loadNamespace("XR")
<environment: namespace:XR>
> eligibleMethods(XR::asServerObject)
Method tags for function "asServerObject" (package XR)
[1] "ANY"              "AssignedProxy"    "ProxyClassObject"
[4] "ProxyFunction"    "name"
> eligibleMethods(XR::asRObject)
Method tags for function "asRObject" (package XR)
[1] "ANY"           "ProxyObject" "list"           "vector_R"
```

In this chapter, we're not concerned with what the methods do, but with what happens when other packages extend XR.

The XRPython package works essentially with the conversion mechanism inherited from XR. The data model in Python is not that different from JSON.

```
> loadNamespace("XRPython")
<environment: namespace:XRPython>
> eligibleMethods(XR::asServerObject, package = "XRPython")
Method tags for function "asServerObject" (package XR)
Methods from package: XRPython
[1] "XMLInternalDocument#PythonObject"
> eligibleMethods(XR::asRObject, package = "XRPython")
Method tags for function "asRObject" (package XR)
Methods from package: XRPython
NONE>
```

Actually, there would be no methods at all, except for a method to convert certain XML objects in R. This implements the method suggested in Chapter 13 (page 297).[2] We don't care why, but the point to note is that XRPython can simply

---

[2]More realistically, the conversion method would be in an application package, such as shakespeare, that needed such objects in Python.

add this one special method, leaving the rest of the method selection as it was by default.

For the XRJulia package the picture is very different. In this case the decision was to replace the JSON mechanism with one more suited to Julia, which has much closer resemblance to R.

```
> loadNamespace("XRJulia")
<environment: namespace:XRJulia>
> eligibleMethods(XR::asServerObject, package = "XRJulia")
Method tags for function "asServerObject" (package XR)
Methods from package: XRJulia
[1] "ANY#JuliaObject"
[2] "AssignedProxy#JuliaObject"
[3] "ProxyClassObject#JuliaObject"
[4] "ProxyFunction#JuliaObject"
[5] "array#JuliaObject"
[6] "list#JuliaObject"
[7] "name#JuliaObject"
> eligibleMethods(XR::asRObject, package = "XRJulia")
Method tags for function "asRObject" (package XR)
Methods from package: XRJulia
[1] "vector_R#JuliaInterface"
```

As we'll see when we look at the selection mechanism, the key distinction is that XRJulia replaced the default `asServerObject()` method with one specialized to Julia.

## Method selection: the procedure

Let's now consider the general strategy for method selection.

We start with a generic function and a call to that function. The function has a *signature*, a subset of the formal arguments for which methods may be defined. At any time, only some of the arguments may have appeared in method specifications; these constitute the *active* signature. Only the active signature matters for method selection. Unless stated otherwise, that's what we mean by "signature" here.

All methods for this generic function from all packages currently loaded are collected into a table in the generic function. Each method was defined with its signature; namely, the class specified for each of the arguments in the generic's signature. If that argument was not specified, it corresponds to class "ANY". The table stores the methods according to a tag computed from the method's signature. The tag is the names of the classes, pasted together separated by "#". In R:

```
paste(signature, collapse = "#")
```

These tags are what `eligibleMethods()` returned in the examples above. After loading XRJulia, the tags for function `asRObject()` were:

```
> eligibleMethods(XR::asRObject)
Method tags for function "asRObject" (package XR)
[1] "ANY#ANY"                   "ProxyObject#ANY"
[3] "list#ANY"                  "vector_R#ANY"
[5] "vector_R#JuliaInterface"
```

We'll refer to this as the table of *defined* methods. The generic function also has a table of *all* methods selected so far, including those selected by inheritance. We'll refer to both tables in describing method selection.

When the generic function is called, the classes of each of the actual arguments define the signature for which we want a method, referred to as the *target* signature. For example, evaluating an expression in Julia might return a string that, when parsed, produced a `"list"` object. Then `asRObject()` would be called with that object and the evaluator (whose class is `"JuliaInterface"`) as arguments. The tag for the target signature is then:

```
"list#JuliaInterface"
```

If an argument is missing in the call, the corresponding element of the target signature is `"missing"`—not `"ANY"` and not the class of the default expression if that exists and were evaluated.

Here then is the procedure for selecting a method:

1. If the target signature exists in the table of all methods, it is selected and returned.

2. If not, a list of eligible inherited methods is formed. If there is only one, it is selected. If there are none, selection fails.

3. If there are multiple eligible methods, those are compared and the preferred methods are retained. If there is only one, it is selected.

4. Otherwise, selection is ambiguous. The lexically first method is selected but a note is printed explaining the ambiguity.

5. The selected method is added to the table of all methods, assigned with the target signature.

We need to define two terms used here: "eligible" and "preferred". Also, there are additional steps for those functions that belong to a group of generic functions, as in the case of binary operators.

A method with a given defined signature is *eligible* for a call with a target signature if, for each argument in the signature, the defined class is the same as the target class or a superclass of the target class. If the generic function has an associated group generic, then a corresponding method for the group generic is eligible if there is no eligible method for the generic function itself for the same defined signature. If there are generations of group generics, the closest generation with a defined method is chosen.

In step 3 of selection, each eligible method is examined. If there is another method preferred to that one, the method is not retained. Among eligible methods, one is *preferred* to another if for each argument the defined class of the first is at least as close as that of the second, and is closer for at least one argument, where "close" is defined by generations of superclasses. A direct superclass is closer than the superclass of a superclass, and so on. The class `"ANY"` is a superclass of all classes, but farther away than any actual superclass. Class `"missing"` has no superclasses except `"ANY"`.

In the case of generic functions belonging to a group, the table of defined methods contains methods defined for a group generic function, provided it did not already have a method for a function that was a member of that group. So, a method for `"Arith"` would be included in the table for `"+"` for signature `"Matrix#Matrix"`, because there is no method for this signature defined specifically for `"+"`.

Also, in step 3 a method is preferred to one with the same defined signature if the second is for a group of which the first is a member. So a method for `"+"` is preferred to one for `"Arith"` which in turn is preferred to one for `"Ops"`.

For a simple example, consider the `asRObject()` call above. There is no defined method with tag `"list#JuliaInterface"` so on the first call with this signature, no method is found at step 1. There are two eligible methods:

    "list#ANY", "ANY#ANY"

The first of these is preferred to the second, since `"ANY"` is a superclass of `"list"`, so only the first method is retained. Selection is unambiguous.

The selected method will be assigned in the table of all methods with tag `"list#JuliaInterface"`.

Although there can be considerable computing to find an inherited method, the computation only takes place once per target signature. Subsequently, the lookup in the table of all methods takes place early on in evaluating the function call. The code to find a method is not called again.

For a more detailed example, suppose we have a call to the `` `+` `` operator:

```
e1 + e2
```

where `e1` has class `"dgeMatrix"` and `e2` has class `"ngTMatrix"` (both classes defined in package Matrix).

There is no defined method for this signature for `"+"` or for either of its group generics. There are six eligible methods:

```
> eligibleMethods("+", c("dgeMatrix", "ngTMatrix"))
Method tags for function "+" (package base)
Target signature: dgeMatrix#ngTMatrix

Tag:       "ANY#ANY" "Matrix#nsparseMatrix" "Matrix#Matrix"
Function:  "+"        "Arith"                "Arith"

Tag:       "dMatrix#nMatrix" "ANY#Matrix" "Matrix#ANY"
Function:  "Ops"             "Ops"        "Ops"
```

None of the methods including `"ANY"` is retained, since there is always a method with a specific class for that argument. Method `"Matrix#Matrix"` is not retained because `"Matrix#nsparseMatrix"` is preferred to it (`"Matrix"` is a superclass of `"nsparseMatrix"`).

This leaves two preferred methods:

```
"Matrix#nsparseMatrix", "dMatrix#nMatrix"
```

The first of these is selected (arbitrarily) with a note to the user:

```
Note: method with signature "Matrix#nsparseMatrix" chosen
  for function "+", target signature "dgeMatrix#ngTMatrix".
  "dMatrix#nMatrix" would also be valid
```

The tags are produced in R order, cycling through the possibilities for the first argument first and so on. You can see all possible tags by omitting the function argument in a call to `eligibleMethods()`. (In the example above, there are 110 possibilities.)

In some examples, the multiple possibilities are in effect doing the same thing, as in the rJava example. Even if the arbitrary choice turns out to be acceptable for whatever reason, users are likely to be unhappy with the message.

Ambiguities can arise when methods in a package supplement or replace related methods from a package this one imports. The XR packages are of this form: the general methods in pacakge XR are intended to be refined by the specific interface packages. Where the new methods are specialized they usually will be clearly selected. The package XRJulia could add one new method for `asRObject()`

and XRPython one for `asServerObject()` because the other methods, including the default, were retained. For `asServerObject()`, on the other hand, XRJulia used a different computation generally. Crucially, this included a default method corresponding to

> `ANY#JuliaObject`

This produces an ambiguous method selection. All the methods in XR leave the second argument as `"ANY"` in the signature:

```
> eligibleMethods(XR::asServerObject, package = "XR")
Method tags for function "asServerObject" (package XR)
Methods from package: XR
[1] "ANY"            "AssignedProxy"     "ProxyClassObject"
[4] "ProxyFunction"  "name"
```

As a result, all of these except the overall default method are ambiguous:

> `AssignedProxy#ANY`

and `"ANY#JuliaObject"` are each preferred for one argument.

This is not just a formality. The two methods can be expected to do something different, and which is correct can only be decided by examining what the methods actually do.

A suggested strategy is to do just that: examine the methods and then copy in the appropriate choices explicitly. The copying in the package's source says that we have consciously decided which of the two candidate methods is appropriate. In the XRJulia example, the XR methods are dealing with proxy objects or explicit user assignments. For these, XRJulia wants to use the existing methods. The source for the package correspondingly copies:

```
.copyFromXR <- c("AssignedProxy","ProxyClassObject",
                 "ProxyFunction", "name")
for(Class in .copyFromXR)
    setMethod("asServerObject", c(Class, "JuliaObject"),
              selectMethod("asServerObject", Class))
```

If the decision was to use `"ANY#JuliaObject"`, a similar action should be done to set the method to that selection.

## Specialized method selection: `as()`

The method selection process described so far is the implementation of the call to **standardGeneric()**, which does method dispatch for all generic functions. It treats all the arguments in the function's signature essentially the same, examining available methods for the target class and its superclasses.

The selection algorithm in the `Methods` package is slightly more general, allowing inheritance computations to be restricted to only some of the candidate classes. Since there is one application of this generality and since that application is important, this subsection examines the computation.

Consider the function `as()`: it has arguments `object` and `Class` and returns the result of coercing the object to be an object from the class. There is also a corresponding replacement function so that `as()` can appear on the left of an assignment, with the interpretation that it replaces the part of the object corresponding to the class by the right side of the assignment. (In this case the object must have been from a subclass of the specified class.)

In an informal sense, `as()` is clearly part of functional OOP: its behavior must depend on the two classes involved, that of the object and that specified by `Class`. Methods are implied automatically when `Class` is a superclass of `class(object)`. In other situations, there is no implication that one class inherits from the other, but if an object from the second class is explicitly needed, there is a sensible definition of what that object should be. For example, one wants to coerce a numeric vector to a character vector, or a named list to an environment, but in neither of these cases do we imply that one class inherits automatically from the other.

For these purposes, the function `setAs()` is provided so that packages can specify coercion methods for new classes. It defines methods for the generic function `coerce(from, to)`, but no actual calls to that function are allowed. Methods for `coerce()` for defined signatures will be supplied by packages when they are loaded, as they would for any generic function, if the packages contain calls to `setAs()`. The resulting methods table for `coerce()` is used and updated by special computations corresponding to a restricted version of inheritance, for a good reason. Methods can be meaningfully inherited for the first argument but not for the second.

To see why, consider evaluating

```
as(object, "C1")
```

for some class `"C1"`. Suppose class `"C2"` is a superclass of `class(object)` and a method is defined to coerce from class `"C2"` to class `"C1"`. Then a method can be defined for coercing from `class(object)` to `"C2"`. Ignoring other arguments to keep things simple, that method is equivalent to:

```
function(from, to)
    as(as(from, "C1"), "C2")
```

But if "C3", say, was a superclass of "C2", a method for coercing `class(object)` to "C3" is no use to us, since the result will typically not be a valid object from the target class, "C2". Aside from any other problems, "C2" can have slots that are not in "C3", and the "inherited" method would tell us nothing about what values those slots should have.

The code for `as()` therefore makes nonstandard use of the method tables for the generic function `coerce()`. If a direct, non-inherited method for the target signature exists, it is used. Otherwise, the inheritance calculations are used, but with an optional argument to use inheritance only on argument `from`.

### Calling the next method or the generic

Methods for a new class of objects can sometimes be conveniently viewed in terms of how they differ from existing methods. The new class may add some information to the previous class, as our examples of dated vectors did. Or the new class may provide similar information to an existing class, but represented somewhat differently. The new method may need to transform some data to conform to an existing method, or to adjust the result of computing with the existing method, or both. If the adjustments before and after are not difficult, the new method can be programmed conveniently around a call to the existing method.

In these contexts, writing the method to use the existing software is not only easier than writing everything from scratch but often a better strategy as well. Improvements in the other software do not need to be transferred to this method.

Two different situations arise, with two corresponding programming techniques. In the "next method" situation, the idea is to invoke a specific method, the one that would apply if this method had not been written, but with adjustments before and/or after. Alternatively, the technique may be to apply the generic function to some different object, usually derived from the current one. The appropriate method may vary from one call to another, but the adjustments before or after will be the same each time.

The first situation can be handled by the `callNextMethod()` mechanism, which avoids re-selecting a method. The second mechanism is handled by the `callGeneric()` mechanism. The two mechanisms may seem similar but in many applications only one or the other will work.

Some typical examples will illustrate the logic. In the discussion of computations for initializing an object from a class (page 178) we noted that `initialize()` methods should provide for subclasses that add slots. Nearly always this is done by a call to `callNextMethod()`, very often as the last line in the method. This example, and the next, are based on code in the Matrix package.[3] This package

---

[3]With apologies to the authors of the package, I have modified the code a little for

provides many methods for sparse matrices. The virtual class `"sparseVector"` provides methods for representing many specific classes, all having in common a numeric slot `"i"` for the indices of the nonzero elements. This slot is required to be sorted. An `initialize()` method for the class takes care of sorting if the call to `new()` included an index argument. Here's the method:

```
setMethod("initialize",
       "sparseVector",
          function(.Object, i, x, ...) {
              has.x <- !missing(x)
              if(!missing(i)) {
                  .Object@i <- # make sure it's sorted
                      ....
              }
              if(has.x) .Object@x <- x
              callNextMethod(.Object, ...)
          })
```

The method allows the call to supply arguments `i` and `x`. The corresponding slots are assigned with `i` correctly sorted (the details are not shown here), if the arguments were supplied.

Then `callNextMethod()` evaluates the default method for `initialize()`—we know the default method is called because the class `"sparseVector"` has no superclasses. The default method will assign slots corresponding to all the named arguments in `...`, which is essential because the actual class of `.Object` may have other slots that this method knows nothing about. That is the central reason for `callNextMethod()` in initialization: one should always allow for subclasses of the defined signature class that include arbitrary additional slots.

Another useful application is to refine a method to suit a subclass. The heavy work can be done by the inherited method, with suitable touch-up by a new method after calling `callNextMethod()`. Subset extraction or replacement often makes sense in this form.

For an application of `callGeneric()`, consider another method from the Matrix package, this time for the group generic function `Logic()`. In this case, we know that both our objects have a slot `"x"` and that the comparison is just that of the two corresponding slots, so long as the matrices have the same dimensions. If not, the method will throw an error. Here is the method:

```
setMethod("Logic",
          signature(e1="lgeMatrix", e2="lgeMatrix"),
```

---

exposition.

```
function(e1,e2) {
    dimCheck(e1, e2)
    e1@x <- callGeneric(e1@x, e2@x)
    e1
})
```

The use of `callGeneric()` is required when the method is written for a group generic function, such as `Logic()`. An actual call will be to one of the members of the group, in this case the operators `` `&` `` and `` `|` ``. For non-group generic functions, methods can use `callGeneric()` as well, but in this case `Recall()` or the explicit function name could be used instead. Each choice is implemented slightly differently but will normally have the same effect.

With `callGeneric()`, a method will be selected on each call, and may be different from one call to another, even for calls *from* one particular method. In contrast, `callNextMethod()` is defined so that each selected method has a unique next method.

The next method is the first inherited method; in other words, the method that is selected by the R method selection, as described on page 197, but with the current method excluded. If only one argument is involved in method selection and the target class has at most one direct superclass, the method chosen will be that from the first superclass having a method, or else the default method if no explicit method is found. Encapsulated OOP has the same notion, usually referred to as "calling the superclass method".

Even for encapsulated OOP, the selection becomes less obvious if the target class has multiple inheritance, that is, more than one direct superclass. Suppose a target class `"C0"` has direct superclasses `"C11"` and `"C12"`, and the current method is that defined for `"C11"`. The next method selection then considers `"C12"` before the superclasses of `"C11"`. With functional OOP, more than one argument may be involved as well, multiple dispatch in object-oriented programming terminology.

As with other general method selection for functional OOP, there is no abstract way to eliminate possible ambiguity. The R method selection will notify users where an ambiguous selection occurs. It's usually a good idea to resolve such ambiguity by adding methods or by replacing `callNextMethod()` with an explicit computation. In practice, a clean class definition tends to make ambiguous method selection less likely.

Because the selection of the next method can be computed from the current method's signature, it can be stored in the method object to avoid recomputing on each call.

# 10.8   S3 Methods and Classes

If you have used R at all extensively, particularly the older and more basic packages, you have very likely encountered S3 methods, sometimes just referred to as "methods", unmodified. These arose during the "Statistical Models in S" project and were incorporated in R as part of replicating that software. Section 2.6 recounts the history in brief.

I wrote most of the original software for S3 methods, which were useful for their application, in the early 1990s. Only a modest change and extension to the existing S software was required, and the classes and methods could be described compactly.

For much the same reasons, S3 functional methods and the analogous R6 package for encapsulated methods [13] remain an easy-learn, lightweight way to add these techniques. For "programming in the small", where the time spent in planning and implementation is to be kept low, these are plausible choices.

However, when the project includes software of lasting importance, and in particular for extending R, the resources offered by formal OOP (in R and in other languages) become important. My opinion in these cases is simple: Don't create new S3 classes when the design and behavior of the resulting software is a major goal.

There are three main problems with the S3 version.

1.  There are no class definitions. Classes are identified by the attribute assigned in each object, but there is no guarantee that the structure of the objects are uniquely defined by this information. In fact, different "classes" use different mechanisms.

2.  Only the first formal argument of the object is queried when selecting a method; that is, multiple dispatch is not supported. For functional OOP this presents some serious limitations, partly got around by a special hack that considers both arguments to some binary operators.

3.  Neither methods nor generic functions exist as classes of objects, making it difficult to make inferences about them. Some structure has been imposed, for example by the S3method() directive described below, but external to the objects themselves.

Using formal classes and methods in R does not exclude either extending S3 classes by S4 classes nor writing formal methods for functions that currently dispatch S3 methods. We've already seen the latter earlier in the chapter, with methods for the plot() function. Extending S3 classes rather than completely replacing them is sensible to continue using existing functions with S3 methods for objects from the new class.

Occasionally, it may be desirable to create new S3 methods, when an S3 "generic" function is called from another function in an existing package (page 209).

The remainder of this section deals with such techniques, in case you need them, and also outlines the S3 method selection mechanism (page 210) in case you need to understand what that is doing.

## Using **S3** classes in **S4**

S3 classes can appear in formal class definitions as slots and in the signatures in method specifications. The only requirement is that they be created and registered by a call to `setOldClass()`. Usually, this just creates a virtual class of the same name with a slot `".S3Class"` containing the `"class"` attribute of the S3 object (which, remember, can have more than one string). Most but not all the S3 classes appearing in the base and stats packages have been registered by code in the methods package, for example `"Date"`:

```
> getClass("Date")
Virtual Class "Date" [package "methods"]

Slots:

Name:    .S3Class
Class: character

Extends: "oldClass"
```

Note that the formal class definition will be created in and from the package containing the `setOldClass()` call.

All registered S3 classes inherit from `"oldClass"`, which can be used to define methods to deal with arbitrary S3 classes. If the class has inheritance in the S3 style, the class attribute will have multiple elements. The full vector should be supplied as the argument to `setOldClass()`. For example, the class `"terms"` is an S3 class, a subclass of class `"formula"` in the S3 sense. To register it:

```
> setOldClass(c("terms", "formula"))
> extends("terms")
[1] "terms"    "formula"  "oldClass"
```

All the specified class names will be registered if necessary (`"formula"` is already registered).

It is possible to have an actual, non-virtual definition created by `setOldClass()` by supplying a formal class definition or the string name of a formal class as the `S4Class=` argument. That is what was done to create the formal version of the `"data.frame"` class shown in Section 10.2, page 148.

As long as the S3 computations conform to the implications of the S4 class, formal OOP computations can be defined for it. This remains an act of faith, since nothing done in the S3 methods or applications is checked against the formal definition. Also, the technique is only useful for S3 classes that follow the formal paradigm by having a set of attributes consistent with slots. Data frames do, most of the time, but other S3 classes may not. For example, the fitted models such as those with class `"lm"` use lists with standardized names for the elements.

Even data frames need care, because other attributes are sometimes introduced that are not slots. The first step in model fitting is usually to compute a *model frame*—a data frame consisting of the variables implied by the formula and other optional arguments. Model frames never became a class but model frame objects have extra attributes not found in other data frames. Computations using a formal version of `"data.frame"` would need to consider these.

The `"terms"` class is another example. It could have a non-virtual S4 class with slots corresponding to attributes such as `"factors"`. But there are many other, optional attributes for `"terms"` objects. As slots these could be `NULL` but as attributes they simply will not be there, so computations using a formal class would need care.

Functional OOP classes can be defined to extend S3 classes as well as defining slots with S3 class. If the S3 class has a non-virtual definition, that should include an inherited class that defines the type of the object, as is the case with `"data.frame"` (type `"list"`). Otherwise, the subclass will have type `"S4"`, which may cause problems with S3 computations.

Model frames are a plausible example. There is no S3 class but the basic idea could be implemented as an S4 class:

```
setClass("model.frame", slots = list(terms = "terms"),
         contains = "data.frame")
```

Defining a subclass of an S3 class can also be used to get around irregularities in the contents of the S3 class. Instead of formal slots for `"terms"`, for example, one could create a subclass with a regular structure having all the slots needed and a suitable initialization method from a `"terms"` object. Most S3 methods should work for such an object, but for those that did not, one could define a function `as.terms()` that returned the S3 object with `NULL` slots stripped away.

## S3 methods for formal classes

Suppose we have a new class, `"myVar"`, that would be useful as a variable in a `"data.frame"` object. The corresponding `data.frame()` function takes any number of individual arguments as potential variables, but existing software will be unable to interpret an object from the new class. Fortunately, `data.frame()` is programmed in an extensible way. It calls `as.data.frame()` for each candidate column, and `as.data.frame()` is an S3 generic function. The solution then just requires two steps:

- define an S3 method, `as.data.frame.myVar()`, for the S4 class in our package;

- declare that method in the `"NAMESPACE"` file for the package:

    ```
    S3method(as.data.frame, myVar)
    ```

    to register the method, without necessarily exporting it from our package.

A natural-sounding approach through S4 methods will not work in this case. Given an S4 class, a natural way to define a method for an existing function would be to convert the function to an S4 generic and then define the method:

```
setGeneric("as.data.frame")

setMethod(as.data.frame, "myVar", ....)
```

While this creates a generic function *from* the function in the base package, it creates the function *in* the current package. All methods defined for this function, from whatever other package, will be kept together and dispatched when the S4 generic is called. However, the function in the base package remains an S3 generic function. At least in the present version of R, it will not dispatch S4 methods. And it is this version of `as.data.frame()` that is, and should be, called from the `data.frame()` function in the same package. Therefore, one should ensure that an S3 method is defined.

Given some care in such situations, recent versions of R will dispatch S3 and S4 methods consistently depending on the objects encountered. S4 method dispatch will understand S3 class structure, provided the S3 classes have been registered by calls to `setOldClass()` and S3 dispatch will understand S4 inheritance when encountering an S4 object.

## S3 method selection

In addition to situations where S3 methods and/or classes are combined with formal OOP, many existing computations are developed entirely with S3 software. In case you need to understand S3 method selection to use such computations in extending R, here is a summary of the method selection mechanism.

S3 method selection and dispatch is a call to `UseMethod()`, with the character string name of the generic as argument. For example function `split()` in the base package is defined as:

```
split <- function(x, f, drop = FALSE, ...)
    UseMethod("split")
```

Functions acting as S3 generics have no special class and so no package information. Two functions of the same name on different packages could not have distinct methods.

The internal code for `UseMethod()` examines the class attribute of the actual argument corresponding to the first formal argument of the generic, `x` in this case (evaluating that argument if it has not been previously evaluated).

A valid S3 class attribute, if not `NULL`, is a character vector. Eligible methods are function objects assigned with the name of the generic followed by `"."` followed by one of the strings in the class attribute. Method selection picks the first such function name among the set of available methods for this function; if none is found, a function corresponding to class `"deflt"` is required and used. Suppose we want to split some date/time data by the different dates:

```
split(z, as.Date(z))
```

where `z` is of S3 class `"POSIXct"`:

```
> class(z)
[1] "POSIXct" "POSIXt"
```

The candidates would be `split.POSIXct()`, `split.POSIXt()` and `split.deflt()`, in that order.

The available methods consist of the functions with appropriate names either visible in the environment of the call or registered from one of the corresponding packages by the `S3method()` directive.

The `UseMethod()` call takes place in the environment from which it is called. It is a primitive function and does not create its own environment. In addition, it has some special internal code that causes the *calling* environment to exit when the method call has been evaluated. Thus it is not possible to include any code after the method dispatch, for example to validate the result of the method, in contrast to S4 method dispatch. (Since we're assuming that new S3 generics are not likely to be written, this is not particularly relevant for programming purposes.)

In addition to functions that contain calls to UseMethod(), some but not all of the primitive functions in the base package can select methods. There is no object-based way to determine which of the primitives take methods, but we can construct an empirical test. The function isPrimitiveGeneric() in the XRtools package tries to construct an S3 method for the specified function, if it is a primitive. If the method is then dispatched, we know that the function dispatches S3 methods. The test cannot be foolproof but seems to identify the correct primitives with the version of R current when this is written. (The implementation shows the convenience of constructing R code dynamically.) The example below applies the function to all the primitives in the base package.

```
> allObjects <- objects(baseenv(), all.names = TRUE)
> primitives <- allObjects[
+    sapply(allObjects,
+       function(x) is.primitive(get(x, envir = baseenv())))]
> length(primitives)
[1] 198
> sum(sapply(primitives, isPrimitiveGeneric))
[1] 99
```

Since sum() of a logical vector is the number of TRUE values, we see that about half the primitives take methods.

# Chapter 11

# Encapsulated Object-Oriented Programming

## 11.1 The Structure of Encapsulated OOP

Encapsulated object-oriented programming in R uses similar concepts and provides similar tools to implementations in other languages. Building on the functional, object-based R software does give a different flavor; for example, because of the "Everything that happens is a function call." rule, the underlying computations use the standard function call mechanism, tuned to behave like encapsulated OOP. Nevertheless, anyone with a background in other implementations should be able to get started without much difficulty. This section outlines the structure; Section 11.2 discusses programming with existing classes; the remaining sections deal with defining classes and with the fields and methods for such classes.

Encapsulated OOP classes in R are designated *reference* classes, to distinguish them from classes in functional OOP. The OOP behavior of objects from reference classes is encapsulated in the class definition, both the properties that the objects can have and the methods that can be invoked on them. As in other encapsulated OOP implementations, properties are extracted, assigned or modified by operating on the object. Methods are invoked on the object by a standard operator.

Objects from reference classes are references, in that modifying a property of an object changes that object in all the contexts where it is currently being used. In contrast to most other R objects, the evaluator will not ensure that changes are local to the function call in which they take place.

Reference classes, like functional classes, are organized within R packages. The combination of a class name and the associated package uniquely identifies the class. The class will typically use functions and other software from its package.

213

Class definitions include the specification of properties (called *fields* in reference classes). Fields have optional typing, and can be declared read-only. Class definitions can inherit from one or more other classes. Methods are defined for the class and stored by name in the class definition. The implementation of a method can refer directly to other methods and to fields. Field and method definitions are inherited: The actual fields and methods in the class definitions will include those inherited, unless direct definitions have overridden the inherited one (usually natural for methods, but dangerous for fields, as we will note).

Like functional classes, a reference class usually has a generator function, often assigned with the name of the class. Generator functions for reference classes have reference-style methods to examine the class and for other operations.

Reference classes in R are by implementation also functional classes: The reference class mechanisms were built on the existing functional class mechanisms. This is to some extent an implementation detail, rather than part of the API for reference classes, but not totally. R remains basically a functional language. Computations on reference class objects will start from some functional expression in R. It is possible to define functional methods for a reference class, but care is needed.

In a reference class for data-frame-like objects, it would be convenient to have subset expressions, like

```
irisx["Sepal.Length",1:50]
```

But if this includes a method for *replacing* similar subsets, the method will not do the same local replacement that a method for functional classes would guarantee. The danger is that functions using the generic subset operators will be invalid if used with a reference class object. We will examine functional methods for encapsulated OOP in Section 11.6, page 235.

If you have used other OOP languages, much of what you have learned should translate to R when using reference classes. In fact, interfacing to other OOP software is one of the key uses for encapsulated OOP in R. There are technical differences, such as using the "$" operator for method invocation rather than the conventional dot. So the method invocation corresponding to `hamlet.findtext()` in Python is just

```
hamlet$findtext("TITLE")
```

in R. If you're learning encapsulated OOP for the first time, on the other hand, experience with the R version should help you later in using other languages. Section 11.2 has some details.

If you have programmed other implementations of encapsulated OOP, programming for reference classes will be somewhat different, partly because the R software is the only case I know in which encapsulated OOP has been implemented on top of functional OOP, in a fundamentally functional language. This has a number of consequences, some superficial and others more fundamental.

Most OOP languages have a separate syntax for class and method definition; in R, all programming takes place in the one functional language. Encapsulated class definitions are created as R objects by calling suitable R functions (outlined in Section 11.3). These functions exist to create an object containing the definition of the class. Largely for reasons of readability, methods are often added to the definition through separate function or method calls. The end result in any case is a complete class definition stored in a meta-data object.

While altering a field has reference-style semantics, the object *in* the field is usually from an ordinary functional class. These objects will behave as they would anywhere in R.

This is the most important distinction resulting from implementing reference classes in R. The distinction is between passing a reference class object as an argument to a function that then modifies one of its fields, versus passing the field itself as the argument. In the first case any modification is seen in the original reference object, in the second case not. This is simply standard R semantics, but programmers need to keep the distinction in mind.

As noted, reference classes are implemented through the existing functional class mechanism in R. The implementation of encapsulated OOP is in fact quite simple, and a relatively thin layer on other programming facilities in the language. The key is that the basic data types in R include a few that behave as references, most importantly the `"environment"` type. Reference classes essentially build an explicit class mechanism on top of environments in R.

# 11.2 Using Encapsulated OOP

The use of classes and methods in R for encapsulated OOP can be described simply. For the experienced user of R, there is no new syntax or operators. Existing expressions are adapted for the purpose; in particular, the `` `$` `` operator is taken over for accessing fields and invoking methods.

A reference class will have a generator object, typically with the same name as the class. Objects from the class can be created by invoking the generator as a function:

```
> wEdit <- dataEdit(data=wTable)
```

If the class does not have a special initialize method, the arguments to the generator are either named fields, such as `"data"` in this example, or unnamed objects from a superclass of this class, in which case all the fields from the superclass object will be assigned the same values in the new object.

Just typing the generator name will, as usual, print the corresponding object via its `show()` method, which in this case gives a brief summary of the class.

```
> dataEdit
Generator for class "dataEdit":

Class fields:

Name:                    data             row.names
Class:         environment data.frameRowLabels

Name:                   edits           description
Class:                   list             character

Class Methods:
     "undo", "edit", "import", "usingMethods", "show",
     "getClass", "untrace", "export", "callSuper", "copy",
     "initFields", "getRefClass", "trace", "field"

Reference Superclasses:
     "dataTable", "envRefClass"
```

The generator has a method for printing self-documentation for the individual method if that was provided:

```
> dataEdit$help("edit")
Call:
$edit(i, var, value)



Replaces the range i in the variable named var in
        the object by value.
```

Once an object from a reference class is created, its methods and fields can be accessed with the `` `$` `` operator. On page 220 we define `"dataEdit"`, a reference class extending another reference class `"dataTable"`. Starting with an object wTable from the `"dataTable"` class:

```
> objects(wTable$data)
 [1] "Clouds"               "Conditions"
 [3] "DewpointF"            "HourlyPrecipIn"
 [5] "Humidity"             "PressureIn"
 [7] "SoftwareType"         "TemperatureF"
 [9] "Time"                 "WindDirection"
[11] "WindDirectionDegrees" "WindSpeedGustMPH"
[13] "WindSpeedMPH"         "dailyrainin"
> wEdit <- dataEdit(wTable)
> wEdit$edit(1:9,"Humidity",NA)
> wEdit$data$Humidity[1:20]
 [1] NA NA NA NA NA NA NA NA NA 94 94 94 94 94 94 94 94 93
[19] 94 94
```

Fields can be modified by the usual R replacement expressions:

```
> wTable$data$Time <- sample(wTable$data$Time)
```

(This shuffles the `"Time"` variable of the `"data"` field, for whatever reason.)

Methods are invoked by the usual syntax, again with the `` `$` `` operator:

```
xx$append(4.5)
```

The difference from similar R expressions for other objects is that the definition of the method can refer to the whole object and potentially alter it, as the `$append()` method does in this example. (To distinguish reference class methods from functions, we will insert the `` `$` `` before the name.)

To understand what happens when using reference classes in R, it's good to remember that these special features are essentially all the basic changes. Reference classes extend R but do not change the language. The various explanations of how things work, such as those in Chapter 3, still apply.

## 11.3 Defining Reference Classes

Fields and methods are the essence of reference classes as of all object-oriented programming. Encapsulated OOP in R is made possible by creating a class with a set of fields and methods defined for it. Each of these will be discussed in a full section, 11.4 for fields and 11.5 for methods. First we give a brief overview of the procedure for defining a reference class.

A reference class is created by a call to `setRefClass()`, much like `setClass()` for functional OOP. Its first argument is the name of the class and a `fields=` argument specifies the fields. The call to `setRefClass()` does two things:

1. assign a definition of the class as a side effect;

2. return a generator object for the class as its value.

Defining reference classes should be done as part of a package for a project that is extending R. The metadata object with the class definition and the generator object will be in the package's namespace. Loading the package installs the class definition.

As an example, here is a definition for class "dataTable". It has two fields, "data" and "row.names":

```
dataTable <- setRefClass("dataTable",
  fields = c(
    data = "environment",
    row.names = "data.frameRowLabels"
  )
)
```

(See page 223 for the motivation.) As with functional classes, the generator returned is a function; calling the function generates a new object from the class. This is the function your users will call to create objects from the class, so assigning the generator object with the name of the class is a good idea in most applications.

As another example, the reference class "Brewery" is defined in the CRAN package Rook [20]:

```
Brewery <- setRefClass(
    "Brewery",
    fields = c("url","root","opt"),
    contains = "Middleware"
)
```

This class has three fields and inherits from class Middleware.

In the definition of "Brewery", the fields "url", etc. were just listed in a character vector, with no requirements about their class. Any R object could be assigned to them, rather as in the Python implementation of OOP. In the "dataTable" example, the fields were the names of a vector whose elements specified the class of that field. The "data" field of "dataTable" will always be of class "environment" or a subclass of that class.

The simple field= argument in the definition of "Brewery" was a short form for:

```
field = c(url = "ANY", root = "ANY", opt = "ANY")
```

There are a number of options in R for specifying the fields. Fields can be made read-only, or defined via accessor methods (the analogue of "setter" and "getter" methods in some OOP languages). All of this will be discussed in Section 11.4.

The essence of encapsulated OOP is that methods are associated with the class rather than with generic functions.

The generator object returned by `setRefClass()` is not a reference object (it couldn't also be a function), but it acts like one in allowing encapsulated-style calls through the `` `$` `` operator. The most important is `$methods()`, which is used to supply method definitions for the class. For example, the `"dataEdit"` class in package XRexamples has methods specified in the package source by:

```
dataEdit$methods(
    edit = function(i, var, value) {
        ....
    },
    undo = function() {
      ....
    })
```

This creates methods named `"edit"` and `"undo"` corresponding to the function definitions in the arguments. Users will call `$edit()` and `$undo()` to operate on objects from the `"dataEdit"` class.

The `$methods()` call to add methods can only appear in the source code for the package creating the class. After that package's namespace is loaded, it is locked to prevent outside code from tampering.

Method definitions can also be included in the `setRefClass()` call itself, but your code will typically be more readable if the method definitions are separate. There can be any number of such method definitions in the source for the package defining the class. It's helpful to separate these and group them by purpose; for example, there will often be a separate definition early in the code for the `$initialize()` method. This is called whenever an object is created from the class; it defines the effect of calls to the class generator itself.

With no arguments, `$methods()` returns the names of all the methods in the class. Similarly, `$fields()` returns the field definitions. The `$lock()` method specifies fields that are to be locked; that is, after the first assignment they will be read-only (an example is on page 224).

## Inheritance for reference classes

In the call to `setRefClass()`, an optional argument `contains=` defines superclasses, whose fields and methods are then inherited by the new class.

For example, here is a reference class for editing objects from the `"dataTable"` class:

```
dataEdit <- setRefClass("dataEdit",
      fields = c(
         edits = "list",
         description = "character"),
      contains = "dataTable")
```

The class is defined with two fields, `"edits"` and `"description"`. It inherits from reference class `"dataTable"`, giving it also the `"data"` and `"row.names"` fields.

The `setRefClass()` call is followed by calls to `$methods()`, such as one creating the `$edit()` and `$undo()` methods. Methods defined for the superclass will be inherited and can be called just like explicit methods. The `"dataTable"` class has a `$initialize()` method, but the `"dataEdit"` class does not. Therefore, a call to the generator for `"dataEdit"` will result in a call to the inherited `$initialize()` method.

If a method wants to invoke the superclass method of the same name, it uses `callSuper()`, giving all the arguments that a direct call would need.

A consideration for inherited reference classes is whether the superclass and subclass are in the same package. In the example above, both classes are in the same package, XRexamples. In contrast, the XRPython package defines a reference class `"PythonInterface"` that inherits from the `"Interface"` class in the XR package.

In the case of different packages, the subclass definition chooses whether objects from the class should inherit from the namespace of the subclass or the superclass. The distinction has implications for methods. Basically, if the superclass has been designed with the intention that subclasses will be defined in other packages, it will arrange for the methods to use only exported objects in its package. In this case, the standard class definition works fine; this is the situation with the `"Interface"` class in the XR package.

If not, the subclass needs to create objects with the same parent environment as the superclass. The optional argument

```
inheritPackage = TRUE
```

in the call to `setRefClass()` arranges for objects to inherit the superclass' package namespace. If methods in the subclass want to use non-exported objects in *their* package's namespace, they need to be written as ordinary functions, as explained in Section 11.5, page 234.

## Designing a class: an example

Before getting into more details, it may be helpful to look at an example of the questions arising when defining a new class. Should the class be functional or reference? What structure (fields, inherited classes) should it have? Also, how to prepare for likely subclasses that inherit from this one?

The example we will look at is the `"dataTable"` class mentioned above. Its fundamental motivation is to provide tools for data-frame like data (observations on a set of variables) at the stage of preparing, correcting and modifying the data.

The data frame was an essential part of the original implementation of functional OOP, implementing the concept of a named list of variables over the same sequence of observations. Along with a model specification, it provided the information needed to fit a model and return the corresponding object. In the original S version [12] and in similar R packages, the computations were naturally functional. The data and the model were arguments, the model-fitting function treated them locally and returned an object of a suitable class. Any modification of the data frame argument would be local to the functions.

This functional approach to data frames is natural in the context and has been widely successful. But other applications of the data frame concept lend themselves more naturally to an encapsulated OOP paradigm and to a reference-based implementation. Perhaps most obviously, the underlying data in most studies doesn't just appear in a perfect form. Much effort goes into preparing data—what is variously called data acquisition, data cleaning and data editing. The process usually involves actual corrections, but often goes beyond glitches or misspecification to questions of what the recorded data means and what form it should have. Different sources corresponding to subsets of the rows of the data frame may need to be adjusted for consistency. Different subsets of variables likewise may need adjusting to be used consistently in the analysis.

All these computations can be done with functional OOP, and many of them have been treated successfully that way in R. However, the basic concept here is exactly matched to the encapsulated OOP model. Suppose we apply the fundamental question (page 133):

> If some of the data in the object has changed, is this still the same object?

The answer is clearly "Yes". The natural paradigm is an object that persists, that may be modified, and that the user wants to examine and perhaps to change.

Let's develop a simple reference class representing this view of data frames. We'll call the new class `"dataTable"`. If you've dealt with data base management software, the `"dataTable"` term suggests a table in such software. There is an analogy, with variables in the statistical sense playing somewhat the role of columns in a database table.

In the original "data.frame" class, the object was a named list, with an additional attribute or slot for row labels. The original S version predated formal classes, but the methods package has an equivalent formal definition:

```
setClass("data.frame",
    slots = c(names = "character",
            row.names = "data.frameRowLabels"),
    contains = "list")
```

In Chapter 10 (page 167) we also looked at a simple alternative, in which the named list was an ordinary slot.

```
setClass("dataFrameNonVector",
  slots = c(
    data = "namedList",
    row.names = "data.frameRowLabels"
  )
)
```

While there is little chance this version will be adopted, it has some design advantages in avoiding inherited behavior that could produce invalid results.

To create a simple reference class for data frames, we can take over the second of the functional classes, but not the first. Reference classes cannot inherit a data part from ordinary R vectors because that would imply that the object had the corresponding type, rather than having effectively type "environment" as all reference class objects do.

Nor would you want that to be the case, because the object would then inherit functional methods that assume changes are local. If a functional method (for fitting a model, say) modified its "local" copy of the object, the original object would change also. One could force automatic conversion of the reference object (essentially by copying it into a functional class object) but that rarely seems a good idea, and certainly not in this case.

Starting then from class "dataFrameNonVector", the most direct analogue of data frames as reference class objects would change the slots into fields. The variables in the "data" field form a named list as before and the "row.names" field again has the class "data.frameRowLabels", meaning that it will be a vector of either integers or character strings.

However, a little reflection suggests a small additional change to be more in the style of reference classes. The "data" field as a named list is a functional class, which is perfectly legal. But the behavior of the field in a reference class would be more consistent if it were itself a reference object. The concept is that data consists of variables referenced by name and that changes to the individual variables will be reflected in the whole object rather than local. So a reference object with named elements would be better.

There is indeed a very natural reference version of such objects—the basic `"environment"` class and data type. With that change to the `"data"` field, we have the class:

```
dataTable <- setRefClass("dataTable",
  fields = c(
    data = "environment",
    row.names = "data.frameRowLabels"
  )
)
```

Keeping the variables in an environment is arguably more natural than in a named list, even aside from the reference question. The numerical order of variables in a data frame is rarely meaningful in itself. It's there, and one is tempted to use it, but typically it represents an artifact of the process by which the data were recorded. Numerical indexing of variables can create obscure and perhaps even dangerous code. Named lists permit any kind of numeric indexing, both from their implementation and from their inheritance from class `"vector"`. There is no simple functional class that allows only indexing by name, and a functional class for data frames could not use a slot of class `"environment"` for the data and retain its functional behavior.

# 11.4 Fields in Reference Classes

This section looks in more detail at how fields are used and designed. The use of fields in reference classes is generally similar to other OOP languages, with the `` `$` `` operator taking the place of the dot character typically used to access a field by name. Fields can have different requirement in terms of content from none at all to a specified class to a customized accessor function.

### Field definitions

All the fields in a class must be named in the corresponding `fields=` argument to `setRefClass()`. It's not required to restrict the contents of the fields; if only the field names are given the contents of the field can be any R object:

```
simpleBox <- setRefClass("simpleBox",
   fields = c("contents", "backup")
)
```

Unrestricted fields are somewhat more efficient to access and, especially, to assign.

As with slots in functional OOP inR, fields can come with a specified class for their contents. The object in the field is then guaranteed to be an object from that class or from a subclass that inherits from it.

```
binaryTree <- setRefClass("binaryTree",
   fields = list(
      left = "treeOrNull",
      right = "treeOrNull",
      leaf = "ANY"
      )
)
```

In this example, `"treeOrNull"` might be the class union:

```
setClassUnion("treeOrNull", c("binaryTree", "NULL"))
```

Some fields can be left unrestricted by supplying their class as `"ANY"`.

Fields may be "locked", in which case they can only be assigned a value once. That will often happen in the initialization method when the object is created, but can take place at any time. Fields contain an `"uninitializedField"` object until they are assigned a value.

The locked fields are specified by the `$lock()` method for a generator function. For example:

```
dataSave <- setRefClass("dataSave",
   fields = c( saved = "dataTable", time = "DateTime")
   )

dataSave$lock(c("saved", "time"))
```

The `$initialize()` method for `"dataSave"` would likely assign the `"saved"` field from an argument and add the current time. After that, no changes in either field would be allowed.

### Field reference

Fields are referenced by name, not by position: there is no concept of a "first" field. Field names should usually be chosen to be syntactically parsed as names, just as for object names. The recommendation is that the names start with a letter and follow with letters, digits and the characters "_" and ".". Fields with a leading "." are legal, but the R implementation uses such names for special purposes, making them dangerous and somewhat misleading otherwise. There is also a mild argument for avoiding "." in field and method names altogether. That character is not legal in names for many other languages; in particular, it is the operator for invoking methods in a number of encapsulated OOP languages, the role of `` `$` `` in R. Your non-R-programming colleagues may find names with embedded dots confusing.

Field names in encapsulated OOP are in fact consistent with the other uses of names in R; that is, they identify objects. As always, the object is specified by the combination of a name and an environment. For encapsulated OOP, the environment is the reference object itself. Both conceptually and in implementation, each object from a reference class corresponds to an environment. Field names identify objects in that environment, either on their own inside the source code for a reference class method or as the right side argument to the `` `$` `` operator.

Field names can appear anywhere in R code on the right of the `` `$` `` operator, with some expression on the left that evaluates to a reference class object. The field name must be one of those specified in the definition of the object's class or in one of the reference classes that are superclasses of that class, as specified by the `contains=` argument in the defining call to `setRefClass()`. All reference objects also have two reserved fields, ".self" and ".refClassDef", corresponding to the object itself and the definition of the class.

The `` `$` `` operator works for any reference object, from any R code. The R methods written for a reference class could use the ".self" field and the `` `$` `` operator to deal with fields. Usually, they do not need to. Instead, fields of the class are available simply by name, both the direct and the inherited fields. Thus, where the "data" field in an object x of class "dataEdit" can be accessed as x$data, when writing a method for the "dataEdit" class, one would simply use the name data by itself. If a reference class inherits from another in a different package, methods may need to be written with explicit use of .self; see Section 11.5, page 234.

In accessing a reference class field or method, the expression on the left of the `` `$` `` operator can be anything arbitrarily complicated as long as it evaluates to an appropriate object. The expression on the right, however, is a constant. It must *be* a name, not just evaluate to one.

If fields must be accessed or changed using a computed name, the reference method $field() can be used. With one argument it returns the field corresponding to the character string value of the argument. With two arguments, it sets the field to the value of the second argument.

## Assigning and modifying fields

The expressions involving fields can also be used to assign or modify data in the fields. These computations are carried out using the standard R computations, with a few special features. We will look at general R computations here. The simplified versions inside method definitions are discussed in Section 11.5.

The various `` `$` `` expressions perform assignment or replacement when they appear on the left of an assignment operator, as they would for any R object. The `"dataEdit"` class has fields `"data"` and `"row.names"`. If eX is an object from that class, the fields can be replaced or modified by expressions such as:

```
eX$row.names <- paste("Obs.",1:nrow)
ex$data$Time <- revisedTime
```

We only need to consider the simple field assignments to understand the general case. As with all R assignments, the replacement expression for `ex$data$Time` modifies a temporary object and re-assigns that object by a simple assignment (Section 5.2, page 74).

When a field is assigned, if the class of the field was specified when the class was defined then the assigned object must come from that class or a subclass. In the definition of class `"dataTable"` on page 223 the `"row.names"` field was specified by:

```
fields = list( data = "environment",
     row.names = "data.frameRowLabels")
```

Any replacement object `value` must satisfy `is(value, "data.frameRowLabels")`.

The field class may be unspecified in which case it defaults to `"ANY"`, allowing any R object. In the `"Brewery"` class on page 218:

```
fields = c("url","root","opt")
```

meaning that all the fields are unspecified as to class.

The definition of the field can be an *access function*. Fields are usually specified in the call to `setRefClass()` by giving the character string name of the required class for the field. If instead the specification of the field is a function of one argument, that function will be called from R when the corresponding field is extracted or modified. The call to the access function has no arguments for extraction and one argument (the new object) for replacement. This is an advanced technique and not recommended for ordinary programming with reference classes. It is used, however, when the reference class object is acting as a proxy for something else. In particular, the technique is central to the XR approach when interfacing R to another language with encapsulated OOP (Chapter 13).

## Reference and non-reference fields

A field in a reference class can be defined to require a particular class, in principle any R class. Whatever the definition, the field itself is a reference. If the class of object x has a field called `"data"`, then whenever the expression x$data appears on the left of an assignment or replacement expression the result is to change the contents of the field everywhere x is used. The same is the case if `data` is replaced by a global assignment (operator `` `<<-` ``) in a method for the reference class.

This is true regardless of the class for the `"data"` field.

On the other hand, if the expression x$data is passed down as an argument in some R function call, what happens depends entirely on the object in the field, and not at all on the fact that we extracted x$data from a reference class object. This just results from standard rules for R computations, but may be something to keep in mind when choosing the definition for a field. There are also some corresponding issues about how much copying gets done, if you are concerned about that (although the differences may be less than users sometimes expect).

A simple example will illustrate. Suppose `data` is some data-frame-like object with named variables, and that we are passing it to an interactive data viewer, such as provided by rggobi. We have some bright idea, never mind what, for a computation that defines some interesting groups of rows. The natural implementation is to provide a function, say `viewWithGroups()`, to compute the Groups variable and add that variable to the data passed on to the viewer:

```
viewWithGroups <- function(frame, ...) {
    frame$Groups <- .... ## some clever computation on the data
    viewer(frame, ...)
}
```

And now `viewWithGroups(data)` gives us the extended view of `data`.

What happens depends only on the class of `data`, regardless of where it came from. For better or worse, the assignment of `frame$Groups` is defined in R both for the functional data type `"list"` (and therefore for classes such as `"data.frame"`) and for the reference type `"environment"` (and therefore for reference classes). In the first case, the computations have no effect on the original `data`; however, in the second case they add a variable or (worse) overwrite one in `data`.

To repeat, this is all just a consequence of the existence of both functional and reference types in R, and of the fact that some computations are valid with more or less the same meaning for two classes from the two paradigms. Object-oriented programming can contribute to safety in these situations. If we really want to modify `frame` only locally, making it an argument of a functional method with signature `"list"`, for example, would protect users.

# 11.5    Methods in Reference Classes

Methods in a reference class are R functions. For the most part they behave as any other function would, but they have two special properties:

- Where they come from: methods are defined and saved as part of the class definition for the reference class, rather than in the namespace of a package as would usually be the case.

- Their environment: while the environment of a function in a package is the namespace of that package, the environment of a reference class method is the object itself (and the parent of that environment is the namespace of the package).

The second property makes the method definition easier to write and clearer to read, by allowing the method to refer by name to fields and other methods. This property is in fact optional. Special circumstances may make it undesirable, in which case methods can be defined as *external*, as will be discussed on page 234. When we talk about reference class methods generally we will mean the standard, internal version.

A reference class method is a function of zero or more arguments, attached to the class definition. The function definitions for reference class methods look like ordinary R functions, and indeed they are with respect to their arguments and to other local computations in the body of the function.

Because the environment of the method is the object itself, fields can be referenced and encapsulated methods called by using the unqualified name inside the method definition. This leads to a programming style similar to languages such as Java or C++, in which field and method names are used directly. In contrast, Python requires fields to be explicitly extracted from a "self" reference to the object. R has a .self field as well, and external methods can be written in the Python style, as shown on page 234. But for all standard method definitions, field and method access only needs the names.

Lets look at a sample method definition to see these and other points. In the "dataEdit" class shown on page 220, the fields are data and edits. The class has two main methods: $edit() and $undo(). They are assigned by a call to $methods() as shown on page 219.

The $edit() method replaces some data in a named variable, recording the change and the previous data in that position, in a list kept in the edits field; $undo() undoes the last edit and pops it off the edits list. The full definition of the $undo() method is:

```
function() {
    "Undoes the last edit() operation
and updates the edits field accordingly.
"
    if(length(edits)) prev <- edits[[length(edits)]]
    else stop("No more edits to undo")
    edit(prev[[1]], prev[[2]], prev[[3]])
    ## trim the edits list
    length(edits) <<- length(edits) - 2
    invisible(prev)
}
```

Some points to note about the definition:

- The body of the definition begins with a multi-line string constant. A string in the first element of a method is interpreted as documentation for the method, a "docstring" in **Python** terminology. This string will be shown in response to a call to the $help() method of the generator. If you use the roxoygen2 package, [38], to write inline documentation for the package, creating the documentation with the `roxygenize()` function will copy the docstring to the regular R documentation for the class.

- The method works by taking the last item in the `edits` field. By design this has the previous values in the edited part of the `data` field, so calling the $edit() method with those arguments undoes the previous edit. Because the environment of the $undo() method is the object, both the `edits` field and the $edit() method are available directly, without using the `` `$` `` operator.

- Finally, the method drops the last two elements from the `edits` field (the original edit and its reverse). This is a *non-local* assignment; indeed, it's the essence of reference classes implementing encapsulated OOP. Therefore the assignment must use the `` `<<-` `` operator.

## Details on writing methods

Assuming you have programmed some R functions and understand the essentials of how these work, reference class methods are reasonably simple, provided you keep a few points in mind.

Other fields of the reference class can be accessed just by using their name, including those fields inherited from reference superclasses. Since fields are not local to the function call, using a field name as an argument name or a local variable masks the field. Don't do it: It's confusing at best and can easily disguise programming errors. The code analysis done when the method is defined will detect and warn about direct misuse.

To *modify* a field, use the R non-local assignment operator `` `<<-` `` as shown in the example. The implementation requires it, but it is also a useful reminder that you are making a reference-type assignment.

As another example, the `$edit()` method in the `"dataEdit"` class has the lines

```
data[i,j] <<- value
edits <<- c(edits, list(backup))
```

These modify and reassign the two fields, `"data"` and `"edits"`.

The object to be assigned can be arbitrary if the field was defined in the call to `setRefClass()` to have class `"ANY"` (which is implicit when no class is given). Otherwise, the object has to be from the declared class, or a subclass. If the field was locked in the class definition by a call to `$lock()`, it can only be assigned once. The assignment may occur in initializing the object, but that is not a requirement.

There may be a lot of available methods, including direct and inherited. A code analysis performed when a method is created detects calls to other methods and ensures that these are made available, by making a reference to them in the environment of the object when created. Other methods in the class or superclasses are not inserted.

For the code analysis to work, there must be some evidence that the other methods are indeed needed. You can guarantee the method's availability by calling the `usingMethod()` utility function in the body of the method. For clarity, put the call before the actual computations, but it does nothing at run time. Its arguments are the names of methods to be used by the present method.

If the other method is called by an ordinary function call, you don't need to declare it via `usingMethod()`. The analysis will detect the call. Also, the `` `$` `` operator automatically includes newly required methods.

But if the other method is used in a hidden way (for example, via a call to one of the `apply()` family of functions, or using `do.call()`), then it is essential to declare the other method via `usingMethod()`. It does no harm to declare methods in any case, and may be a useful reminder.

Inherited methods are available from the `contains=` superclasses specified in the call to `setRefClass()`. These classes will also contain inherited methods, up through the methods defined for class `"envRefClass"`. Methods defined directly override inherited methods with the same name. If there are multiple direct superclasses, duplicate method names are chosen by position. If a class had inheritance defined by `contains = c("C1", "C2")`, then a method from class `"C1"` would be chosen over one of the same name from class `"C2"`. Such duplication is not usually a good sign, though. What should your users expect this method to mean?

If you expect the class being developed to be widely subclassed in other packages, you will do the implementers of those packages a favor if you do *not* call non-exported functions from a method for this reference class. Classes inheriting from your class would otherwise need to take some special action in order to use the inherited methods. For details and workarounds see the discussion of external methods on page 234.

Functions and methods of the same name need to be distinguished. Fairly often, the method name will be the same as a function with a similar purpose, to clarify the computation for users. If you want to call the function from one of the reference class methods, it should be called in the fully qualified form, with the package name.

For example, reference classes will often contain a `$show()` method for automatic printing. If that method wants to call the global `show()` function in the methods package it needs to do so explicitly. In the `"dataEdit"` example, the line:

```
methods::show(data)
```

prints the `"data"` field using the `show()` function, in order to utilize a functional method for that class.

## Debugging reference class methods

The debugging of calls to reference class methods can use essentially all the tools for debugging ordinary function calls. The main point to keep in mind is that the methods are coming from the object itself (as an environment) or from the reference class, not directly from the namespace of a package as would be the case for functions.

Examining results after an error works as usual, for example by setting a suitable `"error"` option:

```
options(error = recover)
```

The stack of calls for browsing will look much as expected and the method definitions for the object will be available from the call; for example, if an error occurred in method `$undo()` in class `"dataEdit"`, the method object `edit` for that class could be examined from a browser in the frame of the `$undo()` call.

The `trace()` and `debug()` mechanisms for debugging before an error occurs both apply, with slight variations that are due to the difference in how the two mechanisms are defined. The `debug()` function and its relatives operate on a function object and can be used directly on a method for an object from a reference class. If `wTable` is an object from class `"dataEdit"`, for example, then

```
debugonce(wTable$edit)
```

will debug the next `$edit()` method invocation on that object.

The `trace()` facility works on names and packages. It has an analogous method `$trace()` that works on either a reference class or an object from the class. For the same object `wTable`,

```
wTable$trace(edit, exit = browser)
```

will invoke the browser on exit from all invocations of `$edit()` on the object.

You can arrange to trace all method calls on all future objects generated from a class by applying `$trace()` to the generator object. To apply the previous trace to all objects, given a generator `dataEdit` for the corresponding class:

```
dataEdit$trace(edit, exit = browser)
```

The same mechanism would not work with `debug()` because the methods for generated objects are copies of the object in the class definition, and `debug()` works on the actual, internal, function in the evaluator.

### Reference class `$initialize()` methods

Reference classes may optionally include a method for `$initialize()`; if so, that method is called when an object is created from the class by invoking the generator function or the `$new()` method. Initialize methods are valuable but also slightly tricky to design well for both functional OOP and reference classes. This section shows some examples and discusses some design hints. Section 10.5, page 178 discusses the analogous functional case.

If no `$initialize()` method is defined for this class or any of its superclasses, the method `$initFields()` is called. This method is defined in class `"envRefClass"`, and takes an arbitrary set of arguments (formal argument `"..."`), passed in from the generating call. Values for fields are supplied as explicitly named arguments. Objects from a superclass, or from this class itself, can be supplied as unnamed arguments; their fields will be copied into the new object, but explicitly named field values override inherited values. For example, suppose `xx` is an object from the class `"dataEdit"` defined on page 220, and that `"dataEdit"` has no `$initialize()` method. Then to create a new object `xCopy` with the same contents but a different `"description"` field:

```
> xCopy <- dataEdit(xx,
+             description = "copy of xx")
```

Note that this is a "shallow" copy; the fields are simply assigned so that reference fields will be the same reference, in this case the same `"environment"` object for `"data"`.

Defining a `$initialize()` method for a class allows more flexible treatment of fields, permits the user's view of the object to be different from the details of the implementation, and may be needed if the class is more than just a reference class (for example, if it is involved with an interface to another OOP language).

Initialize methods can have arguments other than fields or superclass objects. The formal arguments of a reference class method are not determined by a generic function and so are in principal arbitrary. Fields in the class definition will be chosen to be natural for computations with the object. Other formal arguments for `$initialize()` can sometimes be more natural for creating the object.

Two special requirements need to keep in mind:

- Like all methods, those for `$initialize()` will be inherited by any subclasses of this class that are defined, now or in the future. The method should allow for fields other than those in the current definition, so that subclasses will be less frequently required to define their own method. As the example below will show, a natural way to do this is to have the new method call its superclass method, and eventually the default method.

- Arguments to `$initialize()` methods should be optional, in particular so that the generator may be called with no arguments. In the current R implementation, classes are expected to have a default object. For functional classes, these are actual prototype objects; reference classes must create a copy of the prototype, but otherwise they also are expected to generate default objects.

A simple example will illustrate. Class `"dataTable"` has a field, `"data"`, declared of class `"environment"`. To allow users to supply a named list also, an `$initialize()` method might coerce the argument.

```
dataTable$methods(
    initialize = function(..., data) {
        if(missing(data))
            callSuper(...)
        else
            callSuper(..., data = as(data, "environment"))
    })
```

Because argument `data` follows ... in the formal arguments, it can only be supplied by name.

Initializing methods should always `callSuper(...)` to eventually get back to the default, which assigns any named field. Otherwise, any class extending *your* class will have to explicitly initialize all added fields.

## External methods and class inheritance

The function provided as a reference class method is modified so that its parent environment is the object. The modification allows simple reference to other fields and methods, and is slightly more efficient than using the `` `$` `` operator. It is not required, however.

If the first formal argument of the method is `.self`, no special processing is done. In this case the method will be called with the object as the `.self` argument and other arguments being those supplied in the call. No special instructions are required in supplying the method other than the choice of the first argument. Inside this method, the object and its fields and methods are referenced just as they would be in any external function, using the `` `$` `` operator.

Methods of this form, called *external* methods in reference classes, can be included anywhere, but they are most likely in situations where a class in one package extends a class in another package. To see why, suppose one of the methods in our `"dataEdit"` class called a non-exported function in the package where the class is defined, XRexamples.

Now suppose someone defines a subclass, `"dataEdit2"` in another package, newPkg. Standard methods for objects from the new class will have the object itself as their environment, as always. The parent of that environment will be the namespace of a package, and usually of the package in which the class is defined; in the example, the namespace of newPkg. But now the inherited method will fail to find a non-exported function from XRexamples.

When a reference class inherits from a reference class in another package, the subclass definition has the option of including the argument

```
inheritPackage = TRUE
```

in the call to `setRefClass()`. This has the effect of making the parent environment of the object the namespace of the package containing the superclass. In the example, methods for class `"dataEdit2"` would have the namespace of XRexamples as their parent environment, not that of newPkg.

To ensure that both explicitly specified methods and inherited methods work when a class inherits from another package requires some planning in the implementation. Basically, here are the situations:

1. If either class is designed so all standard methods only refer to exported objects, then the other class can do whatever it likes.

2. If any of the superclass methods used non-exported references, then the subclass either has to replace all such methods or else use the superclass namespace by including the `inheritPackage=` argument.

3. If the subclass takes the route of using the superclass namespace, a standard method for the subclass can only refer to exported objects.

Notice that the restrictions are only on standard—i.e., not external—methods. External methods don't use the special code processing for methods and so just behave like any R function in the package.

The main message for projects: if a class being defined is likely to be extended in other packages, it's a good idea to keep its methods free of local stuff. My opinion is that good reference class design will use only exported function calls in method definitions in any case. Aside from the practical issues we have been considering, encapsulated class methods are different from functional methods precisely in that they are included with the objects themselves. In this sense the package is exporting the methods, not just the formal arguments in their calls.

As an actual example, consider the interface packages following the XR structure discussed in Chapters 13 to 15. In this case, the interface design is explicitly to put the common structure into the `"Interface"` class in package XR, but the actual interfaces to specific languages will be from subclasses in separate packages for each language.

The methods for the `"Interface"` class use no non-exported functions in package XR. The requirement was not a big burden. Much of the functionality used was likely to have been called by functions in the specializing packages, and so would have been exported in any case.

# 11.6 Functional Methods for Reference Classes

A call to a generic function, selecting functional methods, may be applied to a reference class object in two contexts:

1. called from an encapsulated method to specialize that method according to the classes of arguments to the method;

2. called directly with the reference class object as one of the arguments.

The first case raises no problems in principle. In effect, it simply says that the definition of the encapsulated method itself depends naturally on the class of one or more objects in addition to the class of the reference object itself. The reference class implementation does not have a mechanism for defining such funcional methods directly, so the encapsulated method has to use a separate generic function for the purpose.

The second case is straightforward in implementation but needs to be approached with caution when the generic function has methods for functional classes as well. There is a danger of invalidating the generic function by implementing non-functional methods.

## Functional methods specializing encapsulated methods

An example from the XR interface classes discussed in Chapters 13 to 15 will illustrate the design options likely to be relevant for this technique.

The XR interfaces use evaluator objects from reference classes to carry out computations in the server language. As part of the process, the object has a method $AsServerObject(). It takes an argument object, an arbitrary R object. The method is required to return a character string such that evaluating the string in the server language returns the equivalent server language object or data corresponding to object. Its definition in class "Interface" is:

```
> XR::Interface()$AsServerObject
Class method definition for method AsServerObject()
function (object, prototype = prototypeObject)
{
    "Given an R object return a string ......"
    asServerObject(object, prototype)
}
```

Other than a documentation string, the entire body is a call to the generic function asServerObject().

Methods for asServerObject() depend on the class of the R object, but potentially also on the reference class. Therefore, the generic function has an argument, in this case prototype, that in effect corresponds to the reference class object. This could often just be passed as .self, but this particular example is a little more subtle: The possibility was allowed that an application might have a particular server language class in mind, so prototype is an optional argument to the reference class method.

In line with the discussion of inherited classes in the previous section, it is desirable for the generic function in such situations to be exported from the package, particularly as the subclass may need to define new methods for it. In the asServerObject() example, a set of methods is defined in XR that could apply for any subclass of "Interface", but some subclasses will choose different methods (in this example, package XRJulia does but XRPython does not).

## Functional OOP for reference class objects

Reference classes are implemented as S4 classes. In principle, the tools of functional OOP are therefore applicable to reference classes, but there are dangers.

R is overall a functional language, meaning that functional computations are available to be applied to reference class objects. To adapt generic functions to reference class objects, defining functional methods for these classes may be attractive, either because an analogous encapsulated method invocation is inconvenient or just to maintain consistent user-level computations. Various operators are examples of the first type. It's more natural to write

```
x + 1
```

than

```
x$add(1)
```

For consistency, one would like to enable the common functions to apply to reference objects when that makes sense:

```
plot(x); summary(x)
```

The danger is that introducing non-functional methods to a functional generic risks invalidating computations using that generic when those computations are unrelated to reference classes.

To avoid potentially disastrous side effects, functional methods for reference classes should abide by the following.

> **Principle:** If a generic function has methods for functional classes, then all methods for this generic should behave functionally.

The motivation here is straightforward and central to the whole idea of functional OOP. A generic function is defined, in functional programming, by describing its arguments and the object it returns. Methods should not violate the computational model that the function represents, but only provide an appropriate *method* for realizing that model.

If the $add()$ method modified x, as is often the case, then it should *not* be replaced by a functional method. To do so risks invalidating computations that assume arithmetic is a purely functional computation.

The principle applies to any non-functional form of data, not just to reference classes. For example, various packages provide mechanisms for dealing with larger objects than R can support internally (see the "Large memory and out-of-memory data" topic in the high-performance task view on CRAN [15]). These mechanisms are frequently not functional, because the data will not be duplicated automatically in the R evaluator. In any case, programmers may want to avoid excessive copying of such large objects. As a result, however, methods must be carefully designed to avoid bad side effects.

An example will illustrate why the principle matters. The **methods** package has a formal class `"structure"` implementing the concept of vector structures. The idea, going back to the original S, is that a vector structure has some data, plus attributes conveying the structure (matrix, time series or whatever). Element-wise operations on the object modify the data but leave the structure unchanged (the `log()` of a matrix is a matrix, for example).

There is a method for the the `Math()` group of generic functions for class `"structure"`:

```
setMethod("Math", "structure", function (x) {
    x@.Data <- callGeneric(x@.Data)
    x
})
```

The method uses the functional OOP implementation of a `".Data"` "slot" that represents the vector part of the object. A nice method, it neatly captures the structure concept; however, it implicitly depends on the functional nature of the objects. The replacement of the `".Data"` slot only affects the object `x` in the function call, not the original argument.

One could imagine a reference class with a `"data"` field that contains a vector of data associated with the object, and various other fields providing additional information. This field acts like the `".Data"` slot and one could write a `Math()` method for the class identical to the `"structure"` method, just replacing

```
x@.Data
```

by

```
x$data
```

The Principle, however, says that this is a bad idea. The `"structure"` method and all similar methods for the `Math()` group generic leave the argument object unchanged. The reference class method modifies it, but should not. The analogous method for the reference class should start with:

```
x <- x$copy()
```

It might be attractive to have a non-functional transformation method, but to avoid potentially disastrous side effects, that needs to use a separate syntax, perhaps

```
x$transform(func,...)
```

to transform the data part of `x` in place by applying the specified function.

Replacement functions are a more subtle case, but the principle applies to these as well. The S-language definition of replacement expressions means that all replacements are equivalent to local assignments. The replacement expression:

```
x[x > upper] <- upper
```

has the same meaning, and is evaluated as:

```
x <- `[<-`(x, x > upper, upper)
```

Methods for the replacement function `` `[<-`() `` can be, and should be functional. The side effect of the replacement is only the implied assignment, and that should be local to the environment where the expression is evaluated.

Legitimate R functions can become non-functional or return wrong results if their arguments have non-functional methods for replacement. It's a plausible strategy to "tweak" some input data in order to produce a desired statistic. The expression above is part of a computation for robust estimation by trimming back extreme values in the input data `x`. The computations are functional and safe in R, assuming the normal object model, but will modify the argument if there is a non-functional method for `` `[<-` ``.

To illustrate the decisions required, let's look again at the `"dataTable"` class. This is a reference class analogue of data frames, with the motivation of data editing and other operations in which changes are made to the object that don't implicitly make it a "different" object, as would be the case for a functional class.

In the process, one will want to try out all or part of the data in its current state with whatever analysis and visualization applies to the project. The objects for these computations should *not* be reference objects: They need to behave as ordinary R objects in the context of the analysis we're trying out. In particular, the Principle says that the functional computations being applied should continue to behave functionally. This says that the data extracted should be from a functional class.

In this example, a different perspective will lead to the same conclusion. The `"dataTable"` class has been created for data editing, but for analysis and functional computing generally, we would rather have a conventional class of data for which many methods are already defined. For `"dataTable"` this is obviously provided by the `"data.frame"` class. Extracting the data in the form of a data frame makes the most existing methods available, as well as being a functional form with no danger of accidentally corrupting the data table.

A simple way to implement the conversion and extraction is to define a method for the `` `[` `` operator that returns a data frame. This will include the expression

x[ ] to return all the data as a data frame. Leaving out a few details of defaults
and error checking, the method is:

```
setMethod("[",
    signature(x = "dataTable"),
    function (x, i, j, ..., drop = TRUE)
    {
        . . . .
        value <- lapply(j,
           function(var) { y <- get(var, envir = x$data); y[i]})
        names(value) <- j
        value <- as.data.frame(value)
        row.names(value) <- x$row.names[i]
        value
    }
)
```

But we do not want a corresponding replacement method. Rather than leaving
the error message to be the obscure:

```
"object of type 'S4' is not subsettable"
```

a more helpful approach is to implement a method with an informative error:

```
setMethod("[<-",
    signature(x = "dataTable"),
    function (x, i, j, ..., value)
    {
        stop(gettextf(
           "No subset replacement for class %s; use $edit()",
           dQuote(class(x))))
    }
)
```

# Part IV
# Interfaces

In implementing extensions to R, the $\boxed{\text{INTERFACE}}$ principle says we should look widely to find good computational techniques to achieve our goals. If an effective solution has been implemented in a form other than R code, providing an interface from R may be the best approach. Part IV of the book looks at how such interfaces can be implemented and made an integral part of an R-based project.

The $\boxed{\text{INTERFACE}}$ principle has always been central to R and to S before. An interface to subroutines was *the* way to extend the first version of S. Subroutine interfaces have continued to be central to R. The approach to them has changed; Chapter 16 discusses current subroutine interfaces, emphasizing an approach that provides convenience and generality through the widely used Rcpp package.

But our options are hugely extended today, because of the immense growth and great diversity of available software. Important software for extending R can come from languages focussing on computations (Python, Julia, C++, $\cdots$), on data organization and management (relational DBs, Excel, $\cdots$) or on display and user interactions (Java, JavaScript, $\cdots$). Chapter 12 reviews interfaces in general, cites some existing packages and discusses concepts for interface programming.

Chapters 13 to 15 present a unified approach to language interfaces, which I refer to as the XR structure. The goals of the approach are convenience, generality and consistency. Application packages can use features of the unified approach to hide the actual interface programming from their users, who program in a natural mix of functions and classes in R. Arbitrary computations and objects in the server language are potential candidates for an interface. The structure is language-independent, with interfaces to a particular language specialized by methods and by functional extensions.

The unified approach is relatively new, having been developed during the writing of the current book. Chapter 13 presents the approach in general. Chapters 14 and 15 describe two interfaces, to the Python and Julia languages. If your interest is specifically in one of these languages, the corresponding chapter can be read independently, referring back to Chapter 13 for details. The packages described in these chapters are available from the Github site `github.com/johnmchambers`.

# Chapter 12

# Understanding Interfaces

## 12.1 Introduction

The ⌐INTERFACE⌐ principle suggests that non-R software should be considered a potential resource for extending R. If there is some suitable software, using that rather than starting over to program something equivalent can save time, and more importantly can improve the quality of the final result.

To have a convenient term, I will refer to the "other language" as a *server language*. This doesn't imply an actual client-server interface, which may or may not be suitable; simply that we view the non-R software as supplying us with something.

Section 12.2 lists some likely languages and existing interface packages for these, with comments on the sort of applications that tend to use each.

In the remainder of the chapter, we discuss various aspects of interfaces and the steps that applications will likely need to take in order to use the server language software effectively.

A basic distinction is between interfaces to individual subroutines and interfaces to other language evaluators (Section 12.3). Chapter 16 describes subroutine interfaces and in particular the Rcpp interface to C++.

For language evaluator interfaces, Sections 12.4 to 12.7 discuss techniques for the inclusion of server language software, for expressing computations, for managing objects and for converting data between the languages.

These sections are also motivation and an introduction to a proposed unified structure for language interfaces. Chapter 13 presents the structure, incorporated in the XR package.

Chapters 14 and 15 present interfaces to the Python and Julia languages using the XR structure.

# 12.2   Available Interfaces

Many languages and programming systems have been used to implement a huge range of computations. The INTERFACE principle encourages us to browse widely. There are many useful forms for the other software, even beyond languages in the usual sense. The common ingredient is some mechanism for programming and carrying out computations that goes beyond the R process and evaluation model discussed in Chapter 3.

Some likely candidates:

**C, Fortran:** These languages were and are the basic implementation languages for S and R. Interfaces to them are still fundamental, and in particular the `.Call()` interface to C is the basic entry point for any software linked into the R process.

**C++:** Programming with C++ functions and classes supports a large body of important algorithmic software. The use of object-oriented structure and some modern programming techniques have produced a general and widely used interface package (Chapter 16).

**Python, Perl, JavaScript, Julia:** These are interactive languages with libraries and capabilities that may be complementary to R. Each provides a general programming environment, in which substantial application software has been implemented, with some tendency to specialize; for example, web-based software in JavaScript, numerical software in Julia.

**Java:** This was traditionally used for serious design of web-based and other graphical interfaces. Its relatively pure OOP structure and thorough facilities for self-describing objects and classes make it natural for similar interfaces from R.

**Haskell:** The most actively used functional programming language.

**Excel, XML, JSON, Relational DBMS:** These languages are particularly important for many projects as repositories for data and, in the case of XML and JSON, as a general mechanism for representing objects to be communicated between languages.

Interfaces to most of these and to other languages have been implemented by many contributors. Table 12.1 lists some R packages providing interfaces.

Our perspective is of interfaces *from* R to another language. This book is about extending R, assuming that one starts from some programming in R, at least in this part of a project. But the ⌐INTERFACE⌐ principle is itself agnostic in this respect. Many approaches to bringing good software together have been valuable.

rpy2 [31] is an interface *to* R from Python that has been widely used. HaskellR [18] is an interesting interface in which R code snippets are inserted into a Haskell program.

Other approaches make multiple languages available from a specialized computing environment. The Jupyter project, `https://jupyter.org`, is a web-based document-creation environment allowing code from Julia, Python, R and other languages to be embedded in a document. The h2o system [25] integrates a range of statistical models with other data-science techniques supporting potentially very large applications with a combination of languages, notably R and Java.

The ⌐INTERFACE⌐ principle suggests that any such approach is worth investigating. With flexible attitudes and well-designed R programming, useful extensions can be adapted to the project at hand.

| Language | Package | |
|----------|---------|---|
| C++ | Rcpp | Chapter 16 |
| Java | rJava | Provides classes, methods |
| Python | rPython, rJython | |
| | XRPython | Chapter 14 |
| JavaScript | V8 | Embedded JavaScript engine |
| Perl | RSPerl | On www.omegahat.net |
| Julia | XRJulia | Chapter 15 |
| JSON | rjson, jsonlite, RJSONIO | Read, write JSON objects |
| XML | XML, others | Other specialized packages |
| Excel | XLConnect | External interface |
| SQL | DBI and its reverse dependencies; RODBC | Relational databases |

Table 12.1: Some packages with interfaces from R to various languages. The boxed entries are discussed in this book; others are available on CRAN unless noted in the table. The upper box of the table has computationally oriented languages, our main focus. The bottom part has data-oriented languages, discussed in Section 12.7.

# 12.3   Subroutines and Evaluators

In the earliest versions of S, an interface to Fortran via the "interface language" was the main technique for extending the language, and was the topic of the book *Extending the S System* [4], whose title suggested that for the present book. The interface language compiled into Fortran subroutines that were linked with the S process.

Today there are many ways to extend R, with programming in R itself being the standard option. But programming in other languages has always remained. The direct descendant of the interface language function is an interface that calls a subroutine compiled from some other language, what we will call a *subroutine interface* for short. Section 5.4 described the various functions in the base package that provide subroutine interfaces.

For serious applications, particularly those dealing with non-trivial structures in the server software, the best way to implement a subroutine interface is usually via the Rcpp interface to C++. For many applications, this will generate a proxy function in R to a C++ function (i.e., to a subroutine returning a value). Conversion between R objects and many C and C++ data types is handled automatically. Programming tools in C++ can extend the conversion capabilities. Admittedly, using Rcpp does require some basic knowledge of C++, but not of any substantial complexity for many applications.

Chapter 16 discusses subroutine interfaces with a focus on using Rcpp. Direct use of the internal interfaces is described in many books on R programming and in Chapter 5 of the *Writing R Extensions* manual.

Interfacing to languages that do not compile separate subroutines requires communicating to some kind of *evaluator* for that language. The rest of the discussion in this chapter and in Chapters 13-15 will be of interfaces to evaluators.

The server language evaluators for an interface differ in whether they are implemented by being *embedded* in the R process or *connected* to it. If the server language is embeddable—if its structure allows it to be loaded and activated inside another process—then there is potential for the R interface to communicate at the C level with the server's evaluator. Many modern languages of a general applicability are potentially embeddable. Java, Python and Julia, for example, are all more-or-less embeddable, as is R itself.

An embedded interface will be defined by a set of C-callable entry points. These will typically include one to initialize the server evaluator. For an eval-parse paradigm, there will be an entry point to pass an expression or command to the evaluator in the form of a text string.

The alternative is to connect the R process to a separately running instance of the server language, via some external communication mechanism. Usually, R writes text to a connection and the server language process then reads, parses and evaluates the text. The interface can pass an arbitrary expression or command directly to the server. Alternatively, the R side of the interface can cast the user's request into a specialized form such as a call to a server-language function specialized to handle requests from R.

Communication through connections is an application of a widely used approach to inter-process communication. The **parallel** package included with R is an example of using socket connections, in this case to communicate among R processes. Connected interfaces are intrinsically best used when they communicate at a fairly high level, rather than through a lot of small-scale computations and data transfer.

The advantages of embedded interfaces are in performance. The function call or other command goes through a C-level call that should have less overhead than inter-process communication. A subtler advantage is that R and the server language are running in the same hardware and software, and so should in principle be compatible as to data representation, text encoding and other environmental parameters. Sometimes the shared environment can be pushed to sharing data in memory as opposed to transforming and copying between the languages.

The advantages of connected interfaces are flexibility and generality. The interface is defined in the two languages themselves, whereas an embedded interface by definition involves the internals of both, increasing the chance that differences between them may complicate portability. By the same token connected interfaces are likely to be more future-proof (being defined in the language) and adaptable (for example, to distributing the processes on multiple machines).

The interfaces to interactive languages in Table 12.1 are mostly embedded. The data-oriented interfaces are mostly organized around writing or reading files.

The XR approach to interfaces in Chapter 13 allows embedded and connected implementations as alternatives for a particular server language, to be selected or even combined according to what suits a particular project or computing environment. The XRPython and XRJulia packages are respectively embedded and connected, as examples, but either could be implemented with the alternate choice.

## 12.4 Server Language Software

In order to interface to a particular function, method or class in a server language, the server software it's based on must be available to the interface evaluator. The different circumstances, which are similar to that for R itself, typically include:

1. Some software will be available by default when the evaluator is started, as are the **base** package and other standard packages in R.

2. Software that must be imported explicitly usually is organized in collections for particular purposes; in R this is the package, but the more common term is *module*, which we'll use here.

3. Modules are frequently organized in libraries; that is, repositories for the module software in which the evaluator will search when asked to import a module. There is often a list of these libraries; in R, `.libPaths()` manages this list.

4. It's likely that an application package using an interface has some server language software of its own, organized in one or more modules. The R objects in the package are available directly from loading the package, but the server language modules and possibly its library must be explicitly made available.

The server languages in Table 12.1 have commands for their version of items 2 and 3, usually a form of `import` command for 2 and either a search list object or a function to manage one for 3. Packages using the XR structure have methods `$import()` and `$addToPath()`.

The application package's own server software can be made available using the location of the installed package.

For example, the **shakespeare** package serving as an example for **XRPython** uses the xml module of **Python** to parse files of data, but also defines its own classes and functions in a file `"thePlay.py"`:

```
class Speech(object):
    . . . .

def getSpeeches(play):
    . . . .
```

(In **Python**, modules are just files of source code.) The **shakespeare** package needs to ensure that this module is available.

Chapter 7 discussed techniques for making non-R code part of a package. The server language needs to be told about the location of the software included. The best general technique is to put server source code in a directory that will be copied to the installed package.

Files and subdirectories of the `"inst"` directory within the source package will be copied to the main directory of the installed package. For example, a collection of **Python** functions, classes and other code could be stored in a directory `"inst/python"` of the application package. The code will be copied to a directory `"python"` of the installed package. **Python** can then be instructed to include the correct directory in its search path. The absolute location of the installed directory of package myPackage is given by the `system.file()` function: in this example:

```
system.file("python", package = "shakespeare")
```

To import modules from our package, the package needs to give Python a directive to add the value of `pythonLocation` to its `"sys.path"` variable. In XR the `$addToPath()` method handles this automatically. For the more basic interface packages, explicit commands can be issued to have a similar effect.

The same `system.file()` function can also give the location of server software in any other available R package. If package pkgA includes some files of Python code in its `"inst/python"` sub-directory of the source directory, the installed version will have this code in the corresponding `"python"` subdirectory. The R function call

```
system.file("python", package = "pkgA")
```

will return the string giving the correct location of this directory.

Modules and other files that are *not* part of an R package will have to be found or specified directly. The details of this unfortunately tend to depend on the operating system and possibly other local variations. The server language itself has a similar problem; in several cases, there is a mechanism similar to R's `system.file()` to get locations in a platform-independent way.

## 12.5  Server Language Computations

Computations from R through an interface are, like all R computations, characterized by the FUNCTION and OBJECT principles: the computation done by the server will be initiated by a function call in R and the information from the server language will be returned as an R object.

The likely server languages differ somewhat in *their* basic computational principles, both compared to R and among themselves. However, three types of computations cover most applications:

1. function calls;

2. invocation of methods on an object and access to its fields;

3. general commands to the evaluator.

Chapter 13 presents a unified structure for interfaces from R to these and other computations. Table 12.2 summarizes some basic interface packages in terms of function calls, method calls and commands.

| Package | Call | Method | Command |
|---|---|---|---|
| rJava | | `.jcall()` | |
| rPython | `python.call()` | `python.method.call()` | `python.exec()` |
| rJython | `jython.call()` | `jython.method.call()` | `jython.exec()` |
| V8[1] | `ct$call()` | | `ct$eval()` |

Table 12.2: Facilities in some interface packages for function calls, method invocation and evaluation of commands. ([1] `ct` = A context object in V8)

The V8 interface to JavaScript differs from the others in that a user creates one or more "contexts" with corresponding context objects. Context objects are similar in many respects to the evaluator objects in XR-based interfaces, but derived from object-based contexts in JavaScript.

In addition to methods, languages supporting encapsulated OOP will have some mechanism for creating named fields within the object. (Whatever they may be called: Java and Julia also call them fields; Python, JavaScript and C++ use "members".) rJava provides an R function, `.jfield()`, to either extract or replace a field. The Python interfaces and the V8 interface to JavaScript require the computation to be done in a command.

Since not all languages follow the equivalent of the [FUNCTION] principle, interfaces often include evaluation of a general command in the server language. The expression for the command is likely to be passed as a character string. Some commands do not have a value and attempting to get one in the server language would be an error; for example, import statements in Python or Julia. The corresponding interface to the command will usually check optionally for successful completion but otherwise return no value.

In the V8 interface, the `$eval()` method accepts expressions that produce no value or others that do, and returns a value for the latter. The Python `exec()` calls never produce a value, even if the supplied expression does. The user needs to assign the result and then retrieve it. Java largely avoids general commands, since all interactive programming has to be through methods. There are commands to control the virtual machine; rJava provides some corresponding specialized interface functions.

# 12.6 Server Language Object References

For computations done entirely in R, the $\boxed{\text{FUNCTION}}$ and $\boxed{\text{OBJECT}}$ principles together describe how the computations work: everything happens as a function call, whose arguments are objects and whose result is another object. For computations that map into function or method calls in a server language, the arguments and the result will continue to follow the $\boxed{\text{OBJECT}}$ principle on the R side.

The strategic question for a particular interface is how to relate the R view to the treatment of data and objects in the server language. How should we refer to these? When and how should we try to convert between objects in the two languages?

A theme that ran through Part I of the book was that the concept of a reference is fundamental to computing and has evolved within essentially all programming languages. So all server languages will have references to objects. How should these be provided to R?

In the rJava interface, a reference to an object from an arbitrary Java class can be returned as an object from the R class `"jobjRef"`. For example, to create a new object from the `"Vector"` class in the Java package `"util"`, using the `.jnew()` function:

```
> v <- .jnew("java/util/Vector")
> v@jclass
[1] "java/util/Vector"
```

The proxy object v can be used as the object for invoking a Java method or as an argument in such a method call.

Other interfaces assume that a computed result in the server language will be converted to an equivalent R object when the function or method call returns. This is not always possible and even when it is, multiple conversions are often undesirable for possible loss of information or simply because they do a great deal of unhelpful computation.

To reuse results, the server object may need to be assigned, automatically as in XR or by an explicit command. In rPython, for example,

```
python.exec("hamlet = xml.etree.ElementTree.parse(myFile)")
```

This is simple enough but does require the application package to keep track of names in the server language. Also, many languages recover memory by garbage collection. This may occur anytime needed. Space is recovered by deleting objects that have no current reference. To economize on memory, the application program may want to explicitly de-reference large objects. Python suggests something like:

```
python.exec("hamlet = None")
```

# 12.7   Data Conversion

Data conversion is nearly always part of any use of an interface in the contexts we're considering.

The languages in the top part of Table 12.1 (C++ through Julia) are chosen usually for computation or display. Data in R may need to be converted as arguments to the computation and data in the server language may need to be converted to make use of the results.

The interfaces in the bottom part of the table are themselves data interfaces; that is, they are widely used for storing data of varying structure. Using these from R can either be a goal in itself (the data for the project is in this form) or can be part of a computationally oriented interface (when both R and the server language can read and write data in this form). The data language interface in the second case may be a better tool for special classes of data than the standard data conversion for a computational server language (as in the use of XML in the example of Chapter 14 to communicate between R and Python).

We will look first at data conversion for computational interfaces and then (page 254) at use of the data language interfaces of Table 12.1.

Potential data conversion issues may relate to the following characteristics of objects in R:

**Vector types.** Some server languages have indexable objects with elements of a single type, analogous to R vectors. Java and Julia are among these, both referring to the objects as *arrays*. Other languages, such as Python and JavaScript, have only lists whose elements can be of multiple types. Applications that use one of the interfaces of this form in Table 12.1 will need to provide local coercion on the R side, one way or the other. Some interfaces, V8 for example, provide related options in the function call that converts the object to R.

**Scalar or vector?** Since R has no scalar types, the package using an interface to server languages that do have scalars may need to control what happens with length-one vectors, to ensure that these are not converted to scalars or that they are, according to the specific situation.

**Matrices.** The Fortran-based representation of numeric matrices as a block storing the data by columns (Section 2.1) tends to be shared by software with an interest in numeric computations with matrices. This includes whole languages such as R and Julia and libraries or modules in other languages, including many in C++ and numpy in Python. Data conversion of such objects between languages sharing the representation can be quite direct. There may be some questions of detail; for example, Julia does not keep an explicit dimension field.

The basic matrix representation, if there is one, in other languages is likely to be incompatible with R, but it's also relatively unlikely that extensive matrix computations will be the target of an interface to these representations. If they are, server language software will need to convert to and from the vector part of an R array.

**Missing values.** R is possibly unique in defining a value in each of its basic types to correspond to missing/Not Available. The international floating-point standard has a range of values corresponding to NaN (Not a Number), which includes R's NA for this type. For other types, such as "integer", no value of the corresponding server language basic type will be interpreted as a missing value. One particular situation is that of logical values; in some languages converting logical missing values to a null-like object (e.g., None in Python) works to create a 3-valued logic such as R supports.

**Integer or numeric?** R is rather loose about distinguishing actual integer data from numeric ("double") data that has integral values, tending usually to produce the latter. For example, the two expressions below produce what seems to be the same sequence, and the default computed value for the by= argument seems to be the same as the one supplied, but the type is integer in one case, double in the other.

```
> seq(1,4)
[1] 1 2 3 4
> typeof(seq(1,4))
[1] "integer"
> typeof(seq(1,4, by = 1))
[1] "double"
```

Such distinctions only occasionally matter in R but may be a problem in a server language. Some special action may be needed to ensure the intended type for the server language data; for example, taking care to format integral values with a decimal point—"1.0" rather than "1".

**Dictionaries and the like.** Although they were not part of the early subroutine-level languages, the data types we will call "dictionaries" are nearly universal as basic data types in the other languages of Table 12.1. Dictionaries are collections of other objects indexed by name. The basic operations are extracting or replacing an element of a given name and removing that element. The data types tend to have two other characteristics: the names are unique (you can't append an element of a given name, only assign it) and indexing elements by numeric location is not meaningful.

R has a class, in fact a basic data type, with just these characteristics: `"environment"`. However, traditional applications in S and R generally use named lists instead; that is, `"list"` vectors with a names attribute. A number of popular functions return such an object as their value.

Named lists have neither of the constraints common to dictionaries: indexing numerically works as with any vector and the names attribute may have duplicates (although some of the software constructing data frames makes the names unique). Also, vectors of other types can have a names attribute in essentially the same way.

The packages in Table 12.1 adopt various conventions in what they recognize and how conversions are done. Applications need to check what will happen and make suitable `as()` calls on data converted back to ensure the intended class for the result.

This discussion has been about data conversion with computationally-oriented languages such as those in Table 12.1.

The bottom part of the table lists interfaces with data as the main focus, each representing data in its own relatively general form. Of the four data-oriented entries in the Table, JSON and XML are the most general, with list- or tree-style representations intended to handle a wide range of data. The other two are less general in their data model but make up for this by being very widely used; inescapable, in fact.

Of the two general languages, XML is much the richer and potentially more complicated. For an extensive discussion of XML with R, see the book by Nolan and Temple Lang, [27]. Since both R and most of the computationally-oriented server languages have packages to read and write XML files, these can serve as intermediate storage for suitable applications. An example running through Chapter 14 uses XML in just this way, starting on page 311.

JSON is much simpler than XML, which can be good or bad. It is easy to use as a base for data conversion but needs to be supplemented with additional user intervention, as is done in the V8 package, or embedded in a more general structure, as in the XR package (Section 13.8, page 286).

Excel data has for decades been the omnipresent form for collecting and often for analyzing business data. This can involve what to an outsider seems a mind-numbing collection of special forms, tricks and hacks, as perhaps suggested by the 100-odd functions in the XLConnect package. But underneath most examples the bulk of the data matches well with the R data frame and the variables in the converted data frame often include familiar R objects, such as vectors of date/times. As a result, interfacing to Excel data can be a powerful tool for extending R.

In my personal birdwatching example on page xv of the Preface, I was adapting my guide's special Excel format to read the day's record into R, simplify it and save it in an R class, "BirdWatch", that specializes "data.frame". As explained in the Preface, the trick here is to recognize a few rows (birds) seen, from a very large table, with bird species in the first column. One then enters a brief note and the number seen or heard. The remaining rows are empty, producing NA in the corresponding cells. Simplified a little, here is the code to read and convert the Excel file.

```
getBirdsFromExcel <- function(file) {
    ## read the Excel worksheet into a data frame, dfr
    tbl <- XLConnect::loadWorkbook(file)
    dfr <- XLConnect::readWorksheet(tbl, 1, header = FALSE)
    .... # Get date, location, preamble from rows 1:5
    there <- !(is.na(dfr[,2]) & is.na(dfr[,3]))
    there[1:5] <- FALSE
    birds <- dfr[there, ] # drop species not seen
    row.names(birds) <- NULL
    names(birds) <- c("Species", "Seen/heard", "Notes")
    BirdWatch(birds,
        date = date, location = location, preamble = preamble)
}
```

The first 5 rows are a hack to store other information (preamble, date and location); one extracts that from the hack and then throws away these rows and all the empty rows, computed in the vector **there**. The resulting data frame, **birds**, is then augmented with the other information in the "BirdWatch" object. We can then go on to display that and, who knows, perhaps even analyze the data in the future.

Like spreadsheets, relational databases are widely used for data sources, including large-scale applications. The interfaces provide access to data sources and a way to maintain large amounts of data in a relatively efficient and convenient form. It's feasible to use these interfaces to manage big data from R, including updating, although the majority of applications access the data without making changes.

The DBI package is a uniform interface to many of these systems. We looked at DBI and its use in Section 2.6, page 43.

# 12.8   Interfaces for Performance

In the discussion so far, the motivation for using an interface has been the "pull" of some software that looks helpful but happens to be written in another language. But an application can also come to consider interfacing to another language from the "push" of extending the performance of existing R software: faster or capable of applying to bigger data.

Is this a good idea? The costs and benefits depend on the programming required and the performance gains obtained.

The scenario is that we have some working R software, perhaps an existing package or else a preliminary implementation. The computations work, but we have reason to believe that they will be unacceptably slow, or perhaps not work at all when applied to the data of interest.

Will reprogramming in another language or using other interface mechanisms be sufficiently better? Where the alternative is a simple transfer of the computations to a different language, the relative improvement may be a more or less constant fraction of the computation time for different problem sizes. A commonly blamed source of performance problems is the per-function-call overhead of the R implementation. The `convolve()` function considered in Section 5.5 is as pure an example as possible. The naive R implementation was a double loop that did a small, scalar numeric calculation each time. Function call overhead had to dominate the computing time. An essentially identical implementation in C was faster by a factor on the order of $10^3$ (page 95), which fairly plausibly would hold over a wide range of data sizes.

The same example also showed significant improvements from *vectorizing* the R computations to reduce the number of function calls. However, as pointed out in the example, vectorizing fairly often involves constructing a larger object, in this case an $n$ by $n$ matrix. As problem size increases, the extra storage requirements can be expected to counteract the reduced overhead.

The natural size dependency in challenging applications may grow similarly or even more steeply than quadratic. You need a reasonably good estimate of that dependency before reprogramming the identical computations in an effort to handle bigger data.

Along the same lines, another advantage of doing some thinking and analysis before reprogramming is the chance of finding a better computational method, rather than just reprogramming. In the convolution example, one would in practice use a Fast Fourier transform with computation growing at a sub-quadratic rate, $O(n * log(n))$ at best.

That's the benefit side. In estimating the cost of reprogramming, some preliminary searching and thinking are again strongly recommended. The hope is to find existing software that implements the desired computations or some substantial parts of it, in a medium with much lower overhead. Numerical computations in C++ or Julia, for example, would be promising.

In the end, some programming across the interface will be needed. Here too there are often good choices to be made. Structuring the server language programming similarly to an existing, well-designed R implementation will make the reprogramming easier and mistakes less likely.

Interface software can help; for example, Rcpp has a number of data types and computational operations designed to replicate analogous facilities in R. For some computations, Julia has similarities to R that can be useful.

Pushing the similarity further, some techniques exist to translate computations programmed in R itself into a more efficient computational model. An example is the use of the LLVM software to compile selected R-language code into machine instructions. LLVM, [26], is a general framework for defining mappings from specified language forms to executable instructions. In [33], Duncan Temple Lang describes a project to apply this framework to compile suitable R software into highly efficient computations. In the future, existing R functions may be "compiled" into efficient, independent computational units provided they satisfy some restrictions.

# Chapter 13

# The **XR** Structure for Interfaces

## 13.1   Introduction

This chapter is a reference for the XR interface structure, the basis for interfaces from R such as those to Python and Julia described in the following chapters. The chapter provides details on the structure and its implementation. This section introduces the motivating goals for the design and for the integration of the interface packages into a project for extending R.

I expect that the main use of interfaces will be in writing application packages, where one or more server languages provide useful software for the application. Most readers of this book will, I hope, be involved in developing application packages or other extensions of R.

The chapters on individual interface packages will repeat the information needed to use the XR structure, along with specifics for the language. You can use the interface package without understanding the structure, but as always in this book I feel that understanding helps. That is particularly true with inter-language interfaces, where there are multiple levels on which programming occurs and many options for how to implement solutions.

### Goals

The interface structure aims to support the use of software in one of a general family of languages. For implementing an application using such an interface from R and especially for the users of those applications, the design has three main goals: convenience, generality and consistency.

**convenience:** The application should be able to create functions, classes and other software interfacing to the server language in a natural R form, with the maximum of detail handled automatically. The *user* of the application should see natural R functions and objects, with the interface causing no inconvenience.

**generality:** There should be no intrinsic limitations on the server language functions and objects used. The interface should be able to accomodate arbitrary classes of objects both in R and in the server language, and to communicate these between languages when that is needed.

**consistency:** The basic structure of the interface should be independent of the individual server language where possible, but with suitable extensions and variations to suit that language. This aids both the application programmer and end user by providing a single natural organization.

To support these goals, the XR structure uses a more advanced form of interface than most of the earlier packages discussed in Chapter 12, with facilities that take advantage of the FUNCTION and OBJECT principles in R and of some of the more modern features of most server languages.

## Levels of programming

Extending R using an interface based on this structure involves packages at three levels, with deeper involvement for fewer participants at the deeper levels:

1. application packages, which will import an interface package for one or more server languages in order to use specific software written in those languages;

2. interface packages for individual server languages, providing facilities and programming tools for the application;

3. the XR package, providing the common structure imported and used by the specific interface packages.

The interface software presented by the application packages to their users will nearly all be in the form of proxies and other specialized software that utilizes the facilities of the specific interface packages at the second level. The interface packages will extend classes and use functions from XR. The application package will use language-specific objects with methods and functions sharing the common structure described in this chapter.

The ordinary R user of an application package will be largely or entirely unaffected by the underlying dependance on software from another language. R programming will continue to be functional and object-based, using standard R expressions.

Programming in an application package will use the interface in the same style, regardless of the server language. Differences will arise when languages differ in a relevant way, but otherwise programming will follow the structure described in this chapter, perhaps simplified by convenience functions for the individual server language that still follow the common structure.

The next higher level of involvement would be developing an interface package for a new language or one using a different strategy for a current language, by specializing the XR structure. Each server language package extends classes and provides methods for generic functions defined in XR. The extended classes override methods for particular server language computations, such as evaluating user expressions or returning descriptions of server language classes. The programming required is typically a small fraction of that needed to define an interface from scratch.

In terms of who-has-to-know-what, the result is a desirable "pyramid" shape. A few people will be involved in creating or modifying interfaces to specific languages. They will need to know a moderate number of details about the XR structure.

A considerably larger number will be involved in writing or modifying application packages that make use of such interfaces. I expect many readers of this book to be among them.

A much larger group again will be users of the application packages. A goal of the XR design is that they will typically be able to ignore the interface nature of the software entirely when using and programming with the application package.

As mentioned, however, I believe that *understanding* the XR design will be helpful, even when explicit use of it is not a programming requirement for your project.

# 13.2 The **XR** Interface Structure

The "XR structure" is based on two characteristics assumed for server languages:

1. objects exist that are organized into classes or a similar structure; and

2. expressions are evaluated as function and/or method calls to which the objects are arguments, and which may return objects as their value.

Python, Java, JavaScript, Julia, Perl and other languages follow versions of this paradigm, as does R itself.

The design of the interface structure has these characteristics:

- R programming with the interface uses evaluator objects, from R reference classes, to evaluate expressions and carry out other computations.

- Evaluator objects have methods for evaluation of expressions, data conversion and commonly required programming steps, such as importing modules.

- In addition, there are some generic functions to carry out computations for the evaluator, mostly for data conversion.

- Specialization to particular languages is through a few methods for the evaluator objects and for the generic functions.

- Functional and object-oriented capabilities in the server language map into corresponding proxy functions and classes in R.

- The potential software in the server language is unrestricted in the sense that any function or method in the server language is a candidate, regardless of the class or data type of the arguments or of the value returned, through the use of proxies.

- To communicate to or return from the server an object belonging to any class in R, there is a formalism for representing an object from an arbitrary R class as a named list, or "dictionary" as we will call it. Where it makes sense, a similar formalism can represent an arbitrary server language object.

There will be some limitations on implementing the design, depending on the particular language, but the overall structure seems to apply widely to the languages following this paradigm that are most likely to be attractive for interfaces from R.

Section 13.3 describes the class of evaluator objects, the common superclass of the specific interfaces. This class has methods for the essential interface computations; typically, evaluating an expression, calling a function or invoking a method in the server language and eventually getting information back to R about the results.

Section 13.4 outlines the facilities that will be provided by interface packages to support application programming and some steps needed in organizing the application package. The programming for an application will emphasize creating proxy functions and classes. There will also likely be some facilities for data conversion keyed to the application (e.g., for graphical and other summaries of results).

Section 13.5 discusses the specialization of the general class to specific server languages. A few methods need to be implemented for the specific server language in order to execute basic computations and retrieve language-specific information. Other operations have default definitions but will often be redefined by methods special to the language (to import modules or get information about functions or classes, for example).

The goal of allowing arbitrary computations in the server language is supported by the definition of *proxy objects* in R. Computed results other than simple types of data will be assigned in the server rather than being converted, with the interface call returning to R a key to identify the result, along with the class and size of the corresponding server-language object. Section 13.6 discusses proxy objects.

Proxy functions and proxy classes provide a transparent link to analogous server language computations. An R function that is a proxy will be called by users like any other function, but its sole action will be to pass the call on to its mirror function in the server language. By default, it will use the current evaluator corresponding to its interface class.

A proxy class is similarly an R reference class whose fields and methods are proxies for corresponding server language constructs. Once a proxy for a server class exists, proxy objects returned by the interface with the matching server class will automatically be promoted to objects from the proxy class. Field and method computations on these objects will map into the analogous server computations. The proxy class object contains a reference to the interface evaluator that created the object, so computations with the object do not need to specify an evaluator, even if it is not the default evaluator for that language.

Many server languages will have the ability to return a description of existing functions and classes, as does R. If so, the interface can use that information in constructing the R proxy function or class.

The net result is that typical application package use will create a proxy class object by calling a normal R function and the object created will then be treated as a normal reference class object. Section 13.7 discusses proxy functions and classes.

The goal of representing arbitrary R objects is implemented by a convention for representing the structure of objects in R as a named list containing parts of the class definition plus the representation of the data in the individual object (the slots, attributes or other pieces of the object). Building on a mechanism that can send simple objects between the languages, the representation allows the return from the server of enough information to construct an arbitrary R object. Similarly, an existing R object can be represented in this form in order to send it to server language software that will then extract the relevant information. If the server language supports it, a similar representation of a server language object could be generated (Julia can do this in a very R-like way).

The conversion mechanism can be specialized at several levels by defining methods for generic functions involved in the conversion. This can accomodate features of the server language (for example, exploiting some common data structures in R and Julia). It could also provide for alternative methods for large objects, or use a common intermediate form (for example, XML objects). Section 13.8 describes the conversion mechanism.

# 13.3   Evaluator Objects and Methods

The XR interfaces operate through an *evaluator object*, an R reference class object with methods to carry out computations in the server language, to convert objects between R and the server language and to obtain information such as the definition of classes or functions in the server language. Actual evaluator objects will come from a class defined for the particular language. That class inherits from the "Interface" class described in this section.

The examples in this and the following sections will illustrate direct use of evaluator object methods. These are the base for the interface computations. Keep in mind, however, that this is the low level of interface programming, relevant for specialized needs and for implementing higher-level features like proxy classes.

Users of application packages will likely never see this level. Even the implementers of the application packages will encounter direct methods only when needing to specialize the match between the server language tools and the functions or classes that make sense for the R application. Otherwise, software that creates proxy functions and classes will provide a simpler programming mechanism.

The XR package maintains a table of currently active evaluators, stored by the interface class name. Computations for proxy functions and classes can obtain the current evaluator for their interface class, so that no evaluator needs to be supplied by the user in normal usage. If no evaluator has been started, the request will create one. Users can have as many evaluator objects as they want from any class, but typical applications will only require one. If the evaluator is needed explicitly, a convenience function for the particular interface class will return it (see Section 13.4).

The evaluator methods for computation have a common structure that is central to the goal of unrestricted server language computations. It can be examined by looking at the basic computational method, $Eval().

```
ev$Eval(expr, ...)
```

The `expr` argument is a character string specifying an expression in the server language. The expression is evaluated and the value returned to R:

```
> ev$Eval("1+1")
[1] 2
```

More general computations can incorporate data into the expression, supplied in the "..." arguments. Each argument will be converted to a character string. The requirement is that when that string is parsed and evaluated in the server language, the result is an object or data structure that is "equivalent" to the R object.

```
> ev$Eval("1+%s", pi)
[1] 4.141593
```

The result of evaluating the expression in the server is an (arbitrary) object or data item in that language. The result can always be assigned there, with a proxy object returned from the call to the $Eval() method. As a result, the arguments substituted into an expression as above can be the result of an earlier interface computation, whether it was returned as a proxy object or converted.

The default strategy is to get back converted versions of simple results (typically, scalars); all others are assigned and returned as a proxy. An optional argument, .get=, controls the strategy; the default corresponds to .get=NA, while supplying .get=TRUE will always attempt to get an equivalent R object.

```
> yp <- ev$Eval("%s.lines", speech)
> yp
R Object of class "list_Python", for Python proxy object
Server Class: list; size: 9
> y <- ev$Eval("%s.lines", speech, .get = TRUE)
> class(y); length(y)
[1] "list"
[1] 9
```

(For **speech**, see page 316.) The conversion strategy for sending and getting arbitrary objects is discussed in Section 13.8.

All the proxy functions and methods created by the interface will also have a .get= argument to allow results to be converted when that is desired. As the discussion of the strategy will explain, essentially any object in each language will have *some* conversion to the other that attempts to include all the relevant information in the original object.

This structure allows essentially arbitrary computations in the server language. Suppose, for example, an argument to a server language function requires some object that is not a simple conversion of an R object. One would first evaluate a server language expression to compute that object. This evaluation would return a proxy object reference to R that can then be passed back as the argument in question.

The arguments are incorporated into **expr** via C-style string fields, **"%s"**, and are processed in the interface by using a method, **$AsServerObject()**, that returns a character string asserted to evaluate to the appropriate server language object when parsed and evaluated as part of the expression.

For example, the Python function **parse()** in the "xml.etree.ElementTree" module parses an XML object in a specified file. A call to this function could be generated by:

```
> x <- ev$Eval("xml.etree.ElementTree.parse(%s)", file)
```

If the argument is a proxy object, a server language expression will be substituted that retrieves the corresponding server language object. In most implementations, this is just the "key" in the proxy object that is the name under which the server object was assigned.

For all non-proxy objects, the server language expression represents the R object as a server language expression that will be evaluated. When the object is a scalar number, character string or logical value, the expression is usually the constant representing that value in the server language. In the example, file could be a quoted character string that would be converted into a similar quoted string in Python.

Scalars at the simple end and representation of arbitrary R objects on the general end are the two boundaries for conversion. In between there may be a variety of special cases, depending on the class of the object and/or on the particular server language interface. The conversion strategy in both directions is discussed in Section 13.8.

The same technique for substituting arguments is used for function calls, method invocation and general commands in the server language. Given the name of a function in the server language, a call to it can be carried out by the $Call() method. In the example above:

```
x <- ev$Call("xml.etree.ElementTree.parse", file)
```

In general, ev$Call(fun, ...) calls the server function specified by fun with arguments corresponding to ... and returns the value of the call.

There is an analogous interface method, $MethodCall(), to invoke a method in the server language from arguments specifying the object, the name of the method and any arguments to that method. To invoke the Python method findtext(), with the argument "TITLE", on the object returned above:

```
ev$MethodCall(x, "findtext", "TITLE")
```

Both $Call() and $MethodCall() have a .get= argument to control whether results are kept as proxies or converted. Remember that all three evaluation methods are the base layer, but proxy functions and classes will hide this layer from the end user.

In R every expression has a value, but some languages execute commands that are not expressions, and will throw an exception if treated as producing a value. The evaluator method

```
ev$Command(expr, ...)
```

is like `$Eval()`, but without a return value.

To get a converted version of a previously computed proxy object:

```
ev$Get(expr)
```

To send a converted version of an object to the server and get a proxy back:

```
ev$Send(object)
```

It's generally a useful strategy to explicitly send objects of substantial size or complexity, rather than include them repeatedly as arguments in evaluation.

Some of the methods will not be meaningful for some languages. A language where all programming is essentially via the OOP paradigm (Java, for example), will have little need for `$Call()`. Invoking an encapsulated-style method will not be meaningful for languages not implementing this version of OOP (Julia, for example).

Packages implementing the XR structure for a specific server language will complete and customize the general structure, as will be discussed in Section 13.5 and illustrated by the packages in Chapters 14 and 15.

The XR structure is designed to make the specializations simple, while keeping to our goals for the resulting interface. Evaluator objects for a particular language will come from a subclass of `"Interface"` that provides methods for the actual communication with the language. Specialized data conversion will be implemented by methods for functions `asServerObject()` and `asRObject()`.

## The table of evaluators

To free application programming from managing evaluator objects explicitly, XR maintains a table of interface evaluators currently active. The table stores objects by the class of the evaluator. Each specific server interface defines such a class, with methods that extend or override the `"Interface"` class in XR.

Access to the table is through the `getInterface()` function, typically:

```
getInterface(Class)
```

which returns the current evaluator for the particular class; if no such evaluator is in the table, one will be initialized. With no arguments, the current evaluator (the last one started for any class) is returned. The package implementing a particular server language interface will provide a convenience function that calls `getInterface()` with the right class.

The XRPython package, for example, includes a function, `RPython()`, which will return a reference to an evaluator from class `"PythonInterface"`, the current existing evaluator if there is one or else a newly created one:

```
ev <- RPython()
```

The call to `getInterface()`, and the calls to the specific functions that use it, have additional optional arguments. Use of these will allow the creation of multiple evaluators possibly with different parameters.

Connected interfaces are more likely to use multiple evaluators; embedded interfaces will usually be calling one internal evaluator for the language. With connected interfaces using general socket connections, evaluators on remote servers might be spawned to handle large problems or to use locally available server language modules. The `getInterface()` function has optional `"..."` arguments that will be passed on to the initializer method for the interface class. A special argument `.makeNew=` controls whether to generate a new evaluator. By default, a new evaluator will be started if none exists or if ... arguments are supplied.

In addition, an argument `.select=` can be provided. It should be a function of one argument. That function will be called with a list of the current evaluators of the specified class; it is expected to return an evaluator if one satisfies its requirements, and `NULL` otherwise. A `.select` function could be used, for example, to require an evaluator with a particular host for its connection. It could also be used to cycle through evaluators for distributed computations (that's why it gets the whole list of evaluators as argument).

## 13.4   Application Programming

Evaluator objects provide the essential interface computations. Their methods, fields and other properties allow us to create structure for interfaces and to specialize that to different languages.

For applications, on the other hand, dealing with the evaluator objects explicitly is usually not important. Just having an evaluator there for a particular language is what's needed. Application programming can provide most computations using a "current evaluator" for the particular language, at least by default, freeing the application from generating and managing the evaluator object explicitly.

For most applications, the default evaluator for a particular language is sufficient. A function that uses the evaluator explicitly typically has a formal argument for the evaluator, with the current evaluator for the particular server language as the default. Proxy functions, for example, use this mechanism to hide the evaluator from the user.

The programming required for an application can largely be supplied by three techniques provided by the interface package:

- importing or sourcing server language code;

- defining proxy functions or classes;

- specializing data conversion for particular classes in R or the server language.

We'll deal with the first of these below, the other two in Sections 13.7 and 13.8.

Aside from possible special data conversion requirements, the R programming for the application package typically only requires calling a few functions provided by the interface package, with straightforward arguments.

## Server language programming

Keeping the R side use of the application simple does sometimes depend on a willingness to do some server language programming for the application package. Proxy functions and classes are very simple to specify, but they assume that the server language function or class matches the needs in R. If there is a mismatch with the application's natural computations, the application will need to do some customizing.

The R side could do most or all the customizing, but often it is more natural to augment the server language side. In the shakespeare package, for example, the data structure parsed is in a tree format not natural for examining the underlying text. The package defines a few Python classes that extract sensible data forms and makes proxies for these in R.

During the development of application-specific server language functions or classes, you are likely to feel the need to experiment interactively with the server language code. Making changes or testing different strategies by reinstalling the application package each time is not fun. Usually, the best strategy is to import the relevant server language modules or code into a good interactive development environment for that language. Server languages likely to be attractive for an interface nearly all have such environments, often several (for example, ipython with notebooks for Python; IJulia, Juno for Julia; or the Jupyter multi-language environment).

The XR structure does have an alternative, the $Shell() method. This starts a simple interactive shell that parses and evaluates expressions in the server language. Expressions have to be on one line, unless each line except the last ends in the back-slash continuation character, "\". The shell will have an exit command, typically the corresponding command or function in the server language.

One advantage of $Shell() is that expressions are evaluated in the same context or namespace used by $Eval() or $Command().

## Organizing the application package

The source code for an application package using an interface needs to provide software in R and usually also in one or more server lanaguages. The application package may construct special objects like proxy functions and classes, whose definition itself involves computing with the interface. If you want to distribute the application package, some special organizational structure and steps may be desirable.

The `"inst"` subdirectory should contain all miscellaneous software in a package; that is, anything other than things like R and C-style source, documentation and other items explicitly recognized by the installation procedure. Section 7.2 discussed the procedure by which the files in this directory and its subdirectories are installed with the package.

The XR structure assumes an organization in which the software in a particular server language is in a subdirectory of `"inst"` with the same name as the language: `"python"`, `"julia"`, etc. This convention provides defaults to methods in the interface.

The evaluator will need to have access to the server language software in the application package. For many languages, access requires two steps: updating some form of search path to include directories associated with the package; and importing individual objects or modules into the current evaluator.

The server language will usually have an internal list of directories in which it expects to find software. When an evaluator starts up, the interface package's initialization method adds its directory of server language code to the path. The application package needs to arrange for the same if it has server language functions. A directory is added to the search path of an evaluator by calling the method

```
ev$AddToPath(directory)
```

An application will nearly always want the search path to have this addition for *any* evaluator from this language. Instead of using the method directly, applications will call a language-specific function, such as

```
pythonAddToPath(directory)
```

The result is to add the directory to a table kept by the XR package. When an evaluator is generated, in this case by `RPython()`, all the directories specified in calls to `pythonAddToPath()` will be added.

The functional versions of `AddToPath()` have an extra benefit. If called at installation time from the source of a package, a load action is added that will repeat the call at load time. The search path is augmented both during installation (for proxy class definition, for example) and when the package is loaded.

If the call is from a function in the application package's source code *and* the subdirectory of server language code has been organized by the convention above, the call needs no arguments. Otherwise, arguments `directory` and `package` can be supplied, the latter if the directory is not in this package. In particular, if you need to search a directory not associated with any R package, specify an empty `package` argument; e.g.,

```
juliaAddToPath("/usr/local/juliaStuff", package = "")
```

If the package is empty and the `directory` is not given as an absolute path it is interpreted in the usual R way, relative to the working directory.

The directories added by the interface so far are stored in the field `"serverPath"` of the evaluator:

```
> ev$serverPath
[1] "/Users/jmc/Rlib/XRPython/python"
> ev$AddToPath(package = "shakespeare")
> ev$serverPath
[1] "/Users/jmc/Rlib/XRPython/python"
[2] "/Users/jmc/Rlib/shakespeare/python"
```

The method checks for duplicates, so adding a path twice has no effect.

Within the directories being searched for server language source, the code will be organized into structures, often referred to as *modules*. Details, such as what a module means and the relation between modules and files, tend to be quite language-specific. In some cases there are multiple forms of access (definitely so in R) depending on questions such as whether the object is available by simple name or only fully qualified (the `"package::object"` form in R).

The evaluator method for importing server language code is:

```
ev$Import(module, ...)
```

The first argument is the name of the module from which functions or other objects are to be imported into the evaluator. The remaining arguments modify what is imported and perhaps other options as well, and will vary among interface packages in their interpretation.

As with the search path, applications are likely to want the modules imported for any suitable evaluator. Functional versions of the method, for example,

```
juliaImport(module, ...)
```

will do this, again by using a table of import expressions in the XR package. The current evaluator for the language (if one has been started) and any evaluator started after the call to the function will include the requested import. Repeated instances of the exactly identical import request will only be executed once.

# 13.5   Specializing to the Server Language

In the XR structure, the interface to a particular language will be provided by a package specific to that language (the examples in this book are the XRPython and XRJulia packages for Python and Julia). End users or, more often, implementers of application packages will use functions in those packages to compute via an evaluator object or to define proxy functions and classes. The evaluator objects will be from a subclass of `"Interface"`, for example `"PythonInterface"` in XRPython.

The initialization method for the interface class will establish communication with the server language. The main distinction will be whether the interface is embedded or connected, as described in Section 12.3. An embedded interface will call a C-level entry to the server language to initialize the server language evaluator. A connected interface will create a connection, usually to an external process running the server language. The XRPython and XRJulia interfaces are respectively examples of the embedded and connected approaches.

Once an evaluator exists, the specialization of computations using it depends mostly on `$ServerExpression()` and `$ServerEval()`, two language-dependent methods used in the key step of the `$Eval()` method:

```
value <- ServerEval(ServerExpression(expr, ...), key, .get)
```

The `expr` argument is a string representing a computation in the server language, possibly written by a programmer but more typically generated from one of the other evaluator methods. The string will contain `"%s"` fields, one for each of the `"..."` arguments, which will be replaced by the server language expression for the corresponding argument.

The call to `$ServerExpression()` returns the resulting string. This is passed to `$ServerEval()`, which uses the server side of the interface to parse and evaluate the string. Creating an interface to a particular language turns on customizing these two steps, the conversion to a server language expression and the evaluation of that expression.

Each of the `"..."` arguments to `$ServerExpression()` is converted to a string by calling the generic function:

```
asServerObject(object, prototype)
```

The `object` argument is the element of `"..."`; the `prototype` argument is an object representing the target class for conversion. When `asServerObject()` is called from `$ServerExpression()`, `prototype` is an object that identifies the language.

A package specializing XR to a particular language typically defines methods to specialize `asServerObject()`. Section 13.8 goes into details.

The package specializes the evaluation of a string containing a server language expression by the definition of the reference class method:

```
$ServerEval(expr, key, keepValue)
```

The implementation must satisfy these requirements:

1. The `expr` argument is always a single string, intended to be a syntactically valid expression. The server side of the interface will parse and evaluate that string.

2. The argument `key` is also a string. If it is non-empty, the value of `expr` should be computed and returned in some form. If `key` is `""`, the expression is evaluated, but its value if any is ignored.

3. The `keepValue` argument is always a single logical value. In R this can be `TRUE`, `FALSE` or `NA`:

   **TRUE:** Always assign the value in the server and return an R proxy object.

   **FALSE:** Always convert the value to an equivalent R object and return that.

   **NA:** Return the converted value if simple; otherwise, return a proxy.

4. Regardless of the previous requirements, if the parse or evaluation throws an exception (a *condition* in R terminology), return a corresponding condition object to R.

Most server languages have tools for parsing, evaluation and exception handling; if so, a server language function can be defined to implement `$ServerEval()`. The XRPython package, for example, has a Python function `value_for_R()` with analogous arguments.

If the `key` argument is not the empty string, it can be used to assign the value; in particular, the key is guaranteed to be unique. The XR design does not require that the server return a proxy with the *same* key as passed in; interfaces may do some caching to avoid having multiple references to the same object.

The `keepValue` argument may be `NA`. Server languages do not usually have an `NA` for logicals, but often have some equivalent of `NULL` (e.g., `None` in Python or `nothing` in Julia), which will be the argument value in this case. The server language side of the interface will then use information about the computed result to decide between proxy and conversion.

When `keepValue` is `NA`, the strategy for current interfaces is to convert scalars (single numbers or character strings, typically) while returning other objects as proxies. Any data type known not to have an R equivalent will be kept as a proxy.

Although we use the term "expression" here, not all languages are like R in that everything evaluates to an object. Languages may have "statements" that are syntactically correct and can be executed, but have no value and cannot be nested in other expressions. Empty keys accomodate the evaluation of general statements.

Whether a return value is expected or not, parsing and evaluating the expression may cause the server to throw an exception (a parse or evaluation error, for example). The interface structure in XR accommodates exceptions with some special classes and methods. If an exception occurs, the `$ServerEval()` method is expected to always return an R object from class `"InterfaceCondition"` or a subclass, even if the value was to be ignored otherwise.

The `$ServerEval()` method for a particular server language interface is conceptually a "two-liner":

1. Send the expression as a string to the server language side, along with `key` and `keepNA` in some form, and get the result back as a string;

2. Convert the string result to an R object.

The XR structure has the goal of generality in both languages: being able to evaluate an arbitrary expression in the server and return the result to R; and being able to return an object from an arbitrary R class. These correspond to a general capability in each of the two steps above.

The generality of the first step is provided by proxies in R for objects, functions and classes, discussed in Sections 13.6 and 13.7. To return a general object in the second step, the conversion first interprets the string as an object from one of a few basic classes and then converts that, using a general representation for R objects.

The conversion computation in the second step is shared between code in the server language and in R. As described in Section 13.8, conversion techniques include an explicit representation for an R object from an arbitrary class and, if possible, a similar representation for an arbitrary server language object. This representation opens the way to further customization of the returned value in R.

For example, basic R vectors have a corresponding, but different, representation in Julia. The simple string representation used for returning results, however, has only an untyped list-like vector. To ensure that a vector object of the right type is returned, the server language code generates the explicit representation for an object from class `"vector_R"`:

```
vector_R <- setClass("vector_R",
    slots = c(data = "vector", type = "character",
              missing = "integer"))
```

For this object, slot `"data"` may be a list of the individual elements. A Julia method generates the representation, the corresponding string is turned into an object of class `"vector_R"`, and a method for the generic function `asRObject()` turns that into the desired vector.

Both R and Julia support computations with complex-valued vectors/arrays, but the intermediate representation has nothing corresponding. To return an array of complex data of type "Array{Complex{Float64},1}" from Julia, the server side constructs a dictionary describing a "vector_R" object. Its "data" slot is a list of the values, formatted as strings; its "type" slot is "complex". The "vector_R" method for asRObject() interprets this and produces the intended "complex" vector.

From the application's perspective, the interface should just do the right thing:

```
> x <- ev$Send(1:3 + 1i)
> x
Julia proxy object
Server Class: Array{Complex{Float64},1}; size: 3
> y <- ev$Eval("%s * 0.5",x)
> ev$Get(y)
[1] 0.5+0.5i 1.0+0.5i 1.5+0.5i
```

Section 13.8 describes the conversion of data in both directions, to a string for $ServerExpression() and from a string for the value returned by $ServerEval(). Customization in both directions uses generic functions with methods specialized to the class of the object and/or to the particular interface.

Interface classes will have some additional methods for computations other than basic evaluation.

$ServerRemove(key): Delete the reference in the server implementation previously assigned as key. The goal is to recover memory in the server language, but how this happens or even whether it happens does not affect the interface itself. If no special method is defined, the default version does nothing.

$ServerClassDef(ClassName, ...): For the server language class or data type of the specified name, return an object listing the fields and methods. What is known about the class and what additional "..." arguments are needed will vary with the language. If there is no reflectance information in the language, the method returns NULL. This is the information used in defining proxy classes.

$ServerFunctionDef(what, ...): Return a proxy function for what; that is, an R function object that when called will evaluate a call to the specified server language function. Where possible, the method will use the server object defining the target function, along with knowledge of the calling conventions of the language.

**$ServerSerialize(key, file); $ServerUnserialize(file):**
Serialize the proxy object stored under the key, writing the resulting byte string to the specified `file`; unserialize the contents of the file, returning a proxy object. Used to save proxy objects.

**$ServerGenerator(class, module):** Return the server language expression for the generator function corresponding to the given class and module. The default is to use the class name as the name of the generator and to prepend the module with `"."` as the separator.

**$ServerAddToPath(directory, pos):** Add the `directory` to the language's search path form importing functions, classes, etc. If the `pos` argument is `NA` (the default), append the directory to the path list; otherwise `pos` is the desired position of the new directory, interpreting the position according to the particular server language (but you really need to know what you're doing if you mess with this).

Server language methods for class metadata and for serializing are important for those computations. Nearly all the languages in Table 12.1 have both.

The methods described so far are all used internally, not called by end users or application packages. Some more directly user-visible operations may also have language-specific methods.

The `$Import()` method in the XR package only takes a single module name as argument. R and many other languages provide for importing only some objects from a module. The `$Import()` methods for the interfaces in Chapters 14 and 15 take a module as first argument and additional argument(s), interpreted according to the import semantics of Python or Julia.

Server languages may or may not require the imported name to be preceded by the module name in later expressions. If the function `parse()` has been imported from module JSON, is it defined simply as `"parse"` in the namespace or must it be called as `"JSON.parse"`? The first case implies that a definition of `parse()` previously existing in the server namespace for the interface will either be masked by the imported version or will mask it. The `$Import()` methods in our packages will behave according to the particular language's protocol. This may include options to use either the simple or fully qualified form.

The `$Source()` method is also likely to be replaced by a language-dependent version that will be more efficient, particularly for large files. In Python, modules are interpreted as files, so `$Source()` is a form of `import` command. In Julia, modules and files are distinct; the language has a function, `include()`, that implements `$Source()`.

# 13.6 Proxy Objects

One of the design goals for the XR structure is to support arbitrary server language computations. This requires that any data or object in the server language can be supplied as an argument to `$Eval()` and the other methods described in Section 13.3. Also, the computation can return an arbitrary server language object as its value. *Proxy objects* are essential to provide this generality; that is, objects in R that are essentially a reference to objects in the server language.

If a server language evaluation produces a result that will not be converted to an R object, it assigns the server language object, using a key supplied by the R side of the evaluator. The server language implementation guarantees to retain the object and to find it when the corresponding key is sent from R. On the R side the key is embedded in a proxy object.

If the R proxy object is used as an argument to a subsequent interface method, `$ServerExpression()` substitutes an expression that retrieves the previously assigned object, usually just the key as a name. The assumption is that the object has been assigned in a workspace or module. An interface package that wanted to use some other mechanism for storing the objects (e.g., an explicit table) could do so by providing methods for `asServerObject()`.

The XR interfaces return proxy objects from class `"AssignedProxy"`, a subclass of `"character"` with the data part being the key for the object. The keys themselves are generated by the `$ProxyName()` method in XR, and guaranteed to be unique in a session both within and between evaluators. The specific interface package and the application using it do not need to generate keys for proxy objects, and shouldn't.

The `"AssignedProxy"` class also has slots for the evaluator that created the object, the server class, the module in which the class is defined and the size of the object. The auxiliary information can be useful when deciding whether a computation had the expected result, and potentially in choosing a method to convert the data to R (e.g., to use a special method for converting large objects). The server class and module allow the interface to detect that a proxy class exists for the object. If so, the returned result is promoted to an object from that proxy class, with the `"AssignedProxy"` object as one of the fields.

`"AssignedProxy"` objects are returned from the server language using the general representation of an R object defined in XR (see page 288). 

The strategy for deciding when to return a proxy is that scalars (including single numbers, character strings and logical values) are returned by conversion but other types are returned as proxy objects. The evaluator method `$Get()` forces conversion by calling `$ServerEval()` with a non-empty `key` and `keepValue =` `FALSE`. An interface package could choose a different strategy simply by a different implementation of the $ServerEval() method.

The key returned as the value of a server language expression is not required to be the same as the key passed in from R. Most of the server languages deal with references to objects, meaning that a relatively inexpensive check can be made to see whether the current object is the same as one recently assigned. In this case, it's a good idea to return the previous key. This reduces clutter on the server side and allows R computations to check equality of objects more efficiently.

Proxy objects allow the application to avoid back-and-forth conversion of data. This is desirable for efficiency, but in addition, the object may change subtly on the round trip. Languages may have many similarities—the XR strategy exploits this—but they inevitably have differences, and in particular in terms of the basic structure for data. So, for example, the two R objects:

```
c(1.1, 2.2, 3.3)
list(1.1, 2.2, 3.3)
```

are different, although similar. But the standard Python object corresponding to them is the same:

```
[1.1, 2.2, 3.3]
```

Converting either of the R objects to Python and getting it back will produce the identical result (the list). To maintain the distinction on the server side requires using some special class of objects, as we'll consider in Section 13.8.

Within a session, proxy objects will persist at least until explicitly removed by the $Remove() method and will be unchanged unless by some computation in the evaluator. Across sessions, proxy objects are not preserved, as we consider next.

### Saving proxy objects; serialization

The objects computed and stored in the server language are maintained only while the corresponding evaluator exists and therefore only for the current R session. To save a copy of the objects on a file, evaluators have a $Serialize() method. Essentially all languages of interest for an interface have a mechanism for serialization; that is, for encoding an object as a byte stream that can later be decoded to recover the object, usually with no loss of information. A corresponding $Unserialize() method reads the file and converts the result back to an object.

The XR package has a default implementation in R that converts the object to an R object which is then coded using the R serialize() function. This approach is very much a choice of "last resort". Interface packages for a particular language will instead use the serialization features of that language to encode the server language object without conversion, with a corresponding implementation for unserializing. The interface package will implement methods `$ServerSerialize()` and `$ServerUnserialize()` that use appropriate code in the particular server language for serialize/unserialize computations. The server language implementation is likely to be better on all criteria: less ambiguity or inaccuracy in the encoding and less computation required.

Serialization in many languages is modelled as a multi-object storage. An open output connection is expected as an argument to the serialize function. Repeated calls append more objects; similarly, unserialize methods return each of the objects written to the same file.

For R with its OBJECT principle, it may be more natural to think of a one-to-one relation between file and object. The XR structure supports either view through an **append** argument to the `$Serialize()` method. If **append** is FALSE (the default), the file is opened and truncated; otherwise the next object is appended to the file. Application code that wants to serialize several objects to one file should call the first time with **append=FALSE** to guarantee that anything previously on the file is wiped out.

When using an interface for serialize/unserialize, keeping a connection open between interface calls may be difficult or undesirable. The packages in the XR family support multiple serialization by closing the file and opening it again in append mode. For unserializing, a more restrictive technique of keeping track of the file position between calls would be needed. Instead, the `$Unserialize()` method always reads the entire saved file, returning a list-like object in the server language containing all the serialized objects as elements. `$Unserialize()` has an argument **all**, which should get the same value as used for **append** in the `$Serialize()` call(s) that created the file. If one object is found, the **multiple** argument determines whether the value is the list of length one or the single object.

# 13.7 Proxy Functions and Classes

Just as proxy objects are objects in R that in fact refer to objects in the server language, so proxy functions and classes are used in R but turn into function calls and OOP computations in the server language.

The function objects and the class definitions will preferably be provided in an application package. Users of the application package will call the functions and work with objects from the classes in a standard R way, largely unaffected by the server language implementation.

This section discusses the programming steps to create proxy functions and classes. For the application author, the programming amounts largely to calls to two functions, one that creates a proxy function object and the other that creates an R class to serve as a proxy for a server language class. The actual functions called will likely be specialized for the particular server language, but they all work the same way and in turn will call versions in the XR package, which we describe in this chapter.

The computations to create proxy functions and classes will nearly always use some metadata information in the server language, similar to the function objects and class definition objects in R but with variations reflecting the particular language (page 282). This information may not be easily available when an application package is installed; if not, two mechanisms in the XR package facilities provide alternatives (page 283).

As a running example, we will use the shakespeare package. All calls for proxy function and class creation for this package are collected in the file `"R/proxy.R"`.

## Proxy functions

*Proxy functions* are objects from a subclass of `"function"`. When called as a function, they use the methods of an interface evaluator to call the corresponding server language function and return its result (usually a proxy object). This could be programmed directly just by the appropriate use of the `$Call()` method, but the proxy function has additional information and ensures, for example, that a module is imported if needed. Proxy functions for particular server languages may include more information, such as documentation for the function. For a strongly typed server language, the identity and type of the actual arguments could be checked. In the case of XRPython and XRJulia, most of the checking is left to the server language.

The general class for proxy functions is `"ProxyFunction"` in the XR package. Individual interface packages usually subclass this to include special features. In the XRPython package, for example:

```
PythonFunction(name, module)
```

creates a proxy function object from class `"PythonFunction"`. (The initial capital "P" is because this is a generator for that class and the interfaces have a convention of capitalizing special class names.)

Calls to the proxy function turn into calls to the Python function of the specified name and module. The Python function `parse()` in the xml.etree.ElementTree module can be used from an R proxy by creating an R function:

```
parseXML <- PythonFunction("parse", "xml.etree.ElementTree")
```

Proxy functions provide a more natural and convenient interface than the `$Call()` method. If the application package has this proxy function defined, then

```
hamlet <- ev$Call("xml.etree.ElementTree.parse", "hamlet.xml")
```

is simplified to

```
hamlet <- parseXML("hamlet.xml")
```

A conceptual simplification for the end user is that the evaluator is hidden, being selected automatically in normal circumstances. In addition, the proxy function takes care of importing the module. The user of the application package does not need to call actual interface routines or to initialize the interface evaluator.

## Proxy classes

A *proxy class* defines an R class corresponding to a specified server language class. Objects from the proxy class appear to have fields and methods in R corresponding to those in the server class. Expressions are written in R, usually the same expression that one would write in the server language with the operator `` `$` `` in R replacing the dot operator in the server.

A proxy class in R is created by calling **setProxyClass()** in the XR package, or a server-language-specific function that extends **setProxyClass()**. For example, `"ElementTree"` is a Python class in module `"xml.etree.ElementTree"`. The Python object returned by the call to **parseXML()** in the example above has this class. The package defining and using the proxy function would also likely create a corresponding proxy class by:

```
ETree <- setPythonClass("ElementTree", "xml.etree.ElementTree")
```

(We'll look at this example in more detail in Section 14.5.)

The reference class created is a subclass of `"ProxyClassObject"`. References to its fields and methods are interpreted as proxies for similar field and method access in the server language.

Objects from the proxy class may be created directly in R, but more frequently are the value of function calls or other computations via the interface, as in the call to **parseXML()** in our example.

Methods and fields in the server object can be invoked or accessed from R by standard R OOP expressions, using the names of the method or field in the server class. The `"ElementTree"` Python class has a method **findtext()**, among others. Then the corresponding class in R will also have a method **$findtext()**. The R method call will evaluate a call to the corresponding Python method. For the object **hamlet** in the example:

```
> hamlet$findtext("TITLE")
[1] "The Tragedy of Hamlet, Prince of Denmark"
```

For the user of the application package, this combination of proxy function, class and object achieves the goal of making the computation essentially transparent to its implementation via an interface.

Calls to proxy functions trigger corresponding function calls in the server language. If these return a non-scalar value, the R function will return an assigned proxy object specifying the server class as a field. If the interface evaluator finds a corresponding proxy class, the value returned will be from that class, with the proxy object as a field. The user, or more frequently special computations in the application package, can use the methods and fields of the object as they would be used in the server language.

## Metadata from the server language

In the examples above, the server language function or class was just supplied by name, optionally with the name of the module where it was defined. Nothing more was needed because Python, like most server languages, provides information which can be used to define specific function and class structure. Methods in the interface evaluator inspect the corresponding object in the server language and return the information to R. As in R, the reference to the name and the module (package in R) is sufficient to find the metadata needed for the proxy.

There is a catch, however. The computations to define the proxy function or class will be part of an application package in a typical project to extend R. It may not be feasible to do those computations when the package is installed. To get around such problems, the XR interface structure supports two optional techniques: load actions and/or a setup step. We'll describe these, but first let's examine how the server language metadata is used.

The interface method creating a proxy function object will access the server function as an object, if possible, obtaining the argument names and perhaps types, online documentation or other information.

Similarly for proxy classes, languages will differ in the information available and in how a class can be used. The computation to create the proxy class uses this information to define the fields and methods in R.

The metadata for a proxy function is returned by the $ServerFunctionDef() method, specialized to the individual server language. The generator function for the corresponding proxy function class will use this information; for example, in the initialize() method for class "PythonFunction".

For proxy classes, the metadata for a class definition is obtained from the method $ServerClassDef(), which returns a named list with the fields and methods as elements. The methods component is itself a named list, whose names are the method names, with the elements being the functions to use as proxy methods in R. The default `$ServerClassDef()` in the `"Interface"` class returns `NULL`, indicating that no metadata information is available.

Languages differ in their implementation of classes; variations will be dealt with in `setProxyClass()` and by the definition of `$ServerClassDef()`. For example, Python has no formal fields. The interface attempts to infer the fields using an object from the class, generating the default object from the class unless the `example=` argument supplies another object.

Other languages will behave differently. Julia has composite types, with fields defined by the metadata. Methods are functional, not encapsulated, so the metadata returned by `$ServerClassDef()` does not generate proxy method calls. Java supplies very detailed information on fields and methods, including types for the arguments and for the return value of all methods. A proxy class definition would have unambiguous fields and methods.

## Load actions and setup steps

Load actions are, usually, functions that are called during package loading, with the namespace of the package as the only argument (Section 7.3, page 119).

A setup step is an R script which itself writes software into the source package. The function `packageSetup()` from the XR package is designed to run such setup scripts. `packageSetup()` ensures that the computations have the package namespace available for use and that the environment for the computations is associated with the package (for example, that classes and other OOP objects will have the name of this package in their `"package"` slot).

Load actions and setup steps are relevant for application packages using interfaces if one cannot assume that the server language evaluator and all relevant modules are available when the application package is installed.

Repositories may require installation with only R, the packages supported by the repository and standard compilers available. There will be no problem if the interface is embedded and the modules used are either available through the server language directly *or* supplied in another R package already installed.

If a module or other server language software is actually part of an application package, it is located using `system.file()` to find the installed package, if the package has been installed and the software was supplied in the `"inst"` directory of that package's source directory. This includes a reference to locations in the current package, even if it has not yet been installed, because of the order of actions in the installation process, as outlined in Section 7.2, page 114.

There remains the possibility that either the language is unavailable (and the interface is connected) or that a particular module cannot be assumed to be available at installation time. Moving the computations to a load action will allow the installed version of the package to be created and distributed without additional requirements on the central repository (but *only* if the repository omits the default test load in the INSTALL command).

The actions to define proxy functions and classes can be delayed until load time by defining a suitable load action. For example, in Section 15.5 we define a proxy for the Julia type "SVD". To do this at load time we define a function

```
.loadSVD <- function(ns) {
    genr <- setJuliaClass("SVD", where = ns)
    assign("SVD", genr, envir = ns)
}
```

The namespace argument is used in the call to setJuliaClass() to store the class definition there and in an explicit assignment of the generator object.

With the load action defined, the package's source code needs to set it:

```
setLoadActions(proxySVD = .loadSVD)
```

This assigns metadata in the installed package that is detected during the package loading. The argument name is just to identify the action in any warning or error message.

If a proxy function or class is created at load time, then any R object that depends on it will also have to be delayed. In particular, an R class that has a proxy class as a field or a superclass will need to be created by a load action. The application package will need to do this explicitly.

Suppose an R class "SvdJTimed" is a subclass of the proxy class, with an additional numeric field "time". Suppose it would usually have the definition:

```
SvdJTimed <- setRefClass("SvdJTimed", contains = "SVD",
    fields =c(time = "numeric"))
```

If the definition of "SVD" is delayed until load time, the subclass will need to be defined then also, after the proxy class. This is most easily done by adding the class definition to the load action that was defined for the proxy class.

```
genr <- setRefClass("SvdJTimed", contains = "SVD",
        fields =c(time = "numeric"), where = ns)
assign("SvdJTimed", genr, envir = ns)
```

These two lines would be added to the body of .loadSVD().

An alternative to the `onload` option is to use a setup step to precompute the information. The calls to create the proxy function or class are the same as in a direct call, with an additional `save=` argument.

In the setup step, the `save=` argument is a file name or an open connection. The effect is to write, on that file or connection, an R language call defining the corresponding function or class. The call has all the metadata incorporated explicitly, so evaluating it does not need any server language computations, either at installation or load time.

Here is an example of a setup step, using the shakespeare package as the application and the XRPython interface. The shakespeare package defines proxy functions and classes, some of which refer to the package's own Python code.

The package defines three proxy classes—`"Act"`, `"Scene"` and `"Speech"`—from its own module, thePlay. Their corresponding direct proxy calls would be:

```
Act <- setPythonClass("Act", "thePlay")
Scene <- setPythonClass("Scene", "thePlay")
Speech <- setPythonClass("Speech", "thePlay")
```

Suppose we decide to generate all class definitions in a setup step, with the complete proxy class definitions written onto a file in the package's source directory, say `"proxyClasses.R"` in the `"R"` subdirectory.

It is recommended to put the R script for a setup step in the `"tools"` directory of the source package, as evidence of how the proxy definitions were computed. The setup script in this example, in `"tools/setup.R"`, could be:

```
library(XRPython)
con <- file("R/proxyClasses.R", "w")
setPythonClass("ElementTree", "xml.etree.ElementTree",
               save = con)
setPythonClass("Act", "thePlay", save = con)
setPythonClass("Scene", "thePlay", save = con)
setPythonClass("Speech", "thePlay", save = con)
close(con)
```

The script opens a connection and passes it as the `save=` argument to each of the `setPythonClass()` calls, to write all the output on the same target file.

To carry out the setup step, set the working directory and call `packageSetup()`:

```
>  setwd("~/localGit/shakespeare")
>  XR::packageSetup()
```

We used the default location, `"tools/setup.R"` for the script, so no arguments were needed to `packageSetup()`. The script creates an environment simulating the installation of the package to ensure that `packageName()` returns the correct name. Note that the package must be installed and loadable, *before* running the script. In this example, the module containing the class definitions is imported from the package's directory. The proxy functions and classes don't need to be defined for the initial installation, but if they are to be explicitly exported, the `export()` directives should be inserted after running the stup step the first time.

To use a different name for the proxy object created instead of the server language function or class name, supply an `objName=` argument in the call. The shakespeare package defines a proxy function for the `parse()` function in the Python module dealing with XML data, as shown on page 280. The proxy is renamed as `"parseXML"` to avoid confusion with R's `parse()` function. In a setup script:

```
PythonFunction("parse", "xml.etree.ElementTree",
                    save = TRUE, objName = "parseXML")
```

A similar `objName=` argument to `setPythonClass()` would rename the class generator object. This example does not specify a file name or connection in the `save=` argument. A file with a name constructed from the function name will be written in the package's `"R"` directory.

The setup step would only need to be run again if the server class or function definition changed in a way that affected the script generated.

## 13.8   Data Conversion

The goal of generality in the XR interface strategy seeks to include any computation that can be expressed in the server language. Proxy objects, functions and classes provide the central mechanism for this. The goal also includes the ability to use arbitrary objects in either language, meaning that there must be some way to describe these objects and to convert them to equivalent objects in the other language, as nearly as possible.

This section describes the XR structure for conversion and the procedures for sending data to the server language and getting R objects back.

The actual communication between the languages is through character strings: the `$ServerEval()` method sends a character string to the server language, which parses and evaluates that string and returns the result, also as a character string. The R side of the interface interprets that string as an object.

To send an object to the server, the object is converted to a string by a call to

```
ev$AsServerObject(object)
```

The string returned is an expression in the server language that evaluates to the equivalent of `object`. That string will be substituted into a `"%s"` field, by `$Eval()` or some other method, to produce the eventual argument to `$ServerEval()`.

The `$AsServerObject()` method just calls a generic function:

```
asServerObject(object, prototype)
```

where `prototype`, by default and usually, is an object identifying the interface language. Methods for `asServerObject()` customize the conversion, both in a language-independent way and specialized when the language offers better analogues to R classes (page 292).

To convert a server language result to an R object, the string returned to `$ServerEval()` from the server language is first converted to an object by the function `valueFromServer()`. This object is produced by applying a standard structure for conversion (we'll describe that next). The result is then the argument to a second generic function, `asRObject()` (page 299).

There are three components to the conversion structure: an elementary level that defines string equivalents for a subset of possible objects and two representations defined in terms of the elementary level to specify arbitrary R and server language objects.

The XR package provides an implementation of the conversion structure. Specific interfaces can modify the conversion to a greater or lesser extent. The XRPython and XRJulia packages are examples: the Python interface uses the basic conversion essentially as is; the Julia interface replaces some conversions for classes where Julia has a more direct correspondence to R.

## Elementary object conversion

The XR package implements `asServerObject()` and `valueFromServer()` through the JSON object notation. Objects representable in this notation form the elementary level for conversion. Interface packages that use the JSON implementation only need to supply server-language software to parse the JSON string in order to get objects from R and software to produce a JSON string for the same range of objects to represent the server object returned to R.

JSON (*JavaScript Object Notation*) is a widely used standard for simple data exchange. For our purposes, it defines a representation of an object as a text string for three kinds of objects:

   i. scalars, in the form of numbers (in a version of standard scientific notation), character strings, and the reserved names `true`, `false` and `null`;

  ii. unnamed lists of representable objects, enclosed in square brackets and separated by commas;

iii. named lists of representable objects, with each element represented as a character string name followed by ":" followed by the representation of that element.

Several R packages implement JSON conversions, producing a string from an eligible R object and parsing such a string to return the corresponding object. The current XR implementation uses the jsonlite package, [29]. All the server languages discussed have their own corresponding modules or packages to perform similar conversions.

The second and third forms are referred to in JSON as arrays and objects, but that terminology is too confusing here. Where some general term is needed, the second will be called a JSON list and the third a *dictionary* (the term in Python, Julia and other languages for the analogous structure).

The use of JSON has advantages of simplicity and availability. The standard can be described in a page (`www.json.org`). The notation is widely used; nearly all the languages needed for interfaces have an implementation, and writing one would not be a major undertaking. Several existing basic interface packages use it, including interfaces to Python [6] and JavaScript [28].

The JSON notation is limited in what it can represent. The objects representable in this form do not include all the "basic" R objects, in the sense of the objects that belong to one of the built-in data types (Section 6.1, page 97). Of the vector types, the notation will represent `"logical"`, `"character"` and both `"integer"` and `"double"` forms of numeric. But neither `"complex"` nor `"raw"` can be accommodated. These are handled by the representation for arbitrary R objects.

Also, an R `"list"` object with scalar elements all of the same basic type has the same JSON representation as a vector of that type. The server language may or may not have such a distinction either: Python does not but Julia does.

The natural strategy for an interface package will depend on the server language. Julia has closer analogues to R vectors than does JSON. The XRJulia interface defines conversion methods for these (both in R and in Julia).

Python has an object hierarchy similar to JSON. The XRPython interface uses the default XR structure.

## Representing arbitrary R objects

Dictionaries are the mechanism for representing arbitrary R objects. The dictionary will be interpreted as a general R object if it has an element named `".RClass"`. The value of that element (along with an optional `".package"` element) is taken to specify the class.

There are also reserved names for the type and the inherited classes. Other than these reserved names, the remaining elements of the dictionary will be interpreted as the slots in an S4 class definition, including all inherited slots. A class that extends "vector" or one of the vector types also has a pseudo-slot ".Data" for the data part of the object.

For example, although the "ts" class for time series data is an S3 class, it is consistent with an S4 vector class with a numeric slot "tsp" for the time series parameters. Here is a JSON representation of the uspop data in the datasets package:

```
{ ".RClass" : "ts", ".package" : "",
   ".type" : "double", ".extends" : "ts",
   ".Data" : [3.93,5.31,7.24,9.64,12.9,17.1,23.2,31.4,39.8,
             50.2,62.9,76.0,92.0,105.7,122.8,131.7,151.3,
             179.3,203.2],
   "tsp" : [1790.0,1970.0,0.1] }
```

The representation generalizes to any vector with attributes.

Arrays and their special case, matrices, are included in the XR implementation similarly, although they are not in fact S3 classes. Some special methods in XR give them the appearance of a formal class with slots "dim" and "dimnames", plus a data part. If the interface does not have a special asServerObject() method, a matrix will be converted using the general representation. In the example below, ev is an evaluator for the Python interface:

```
> mProxy <- ev$Send(matrix(1:12,3,4))
> ## Converted in Python to a dictionary:
> ev$Command("print %s.keys()", mProxy)
[u'dim', u'.type', u'.package', u'.RClass', u'.Data', u'.extends']
> ## Getting it back decodes the dictionary:
> ev$Get(mProxy)
     [,1] [,2] [,3] [,4]
[1,]    1    4    7   10
[2,]    2    5    8   11
[3,]    3    6    9   12
```

We printed out the keys (element names) in Python. They include ".Data" for the data part, "dim" and "dimnames" for the slots, plus the specially named elements. Server language computations might interpret this object and produce something more idiomatic for the language, but even working directly from the dictionary, the application could compute some revised values and returned the result to R by this mechanism.

The conversion of basic vectors and other built-in types uses the function `typeToJSON()`, which generates a JSON representation for the data in an object, using only the type and the data. The representation by `typeToJSON()` gives essentially the full information, but not directly for types other than those representable as numbers or character strings. Otherwise, a combination of formatting the individual elements and, as a last resort, serializing the object provides at least a mechanism for transmitting the data and later retrieving it. See page 294 for details.

The slots in an object will often be themselves objects from some class other than the basic data types. The explicit representation will be applied to these as well. For example, in Section 10.2, page 149, we considered the definition of a class `"track"`:

```
track <- setClass("track",
    slots = list(lat = "degree", long = "degree",
                 time = "DateTime"))
```

The slots of this object would be represented in JSON by explicit forms for classes `"degree"` and `"DateTime"`.

R accepts objects from a subclass of the declared class for a slot. For simple inheritance, the server software will still be able to find the required slots. For a virtual class like `"DateTime"`, the R side may need to coerce the slot into a particular representation required by the server computations. In the example above, if the server required the `"POSIXct"` representation of date/times, the R side would require:

```
object@time <- as(object@time, "POSIXct")
```

This could be in application-specific code or in a method for `asServerObject()` for class `"DateTime"`.

Languages can return R vectors of a specific type in the JSON notation by using the `".RClass"` representation for an object from the R class `"vector_R"`. This class represents a vector in an explicit form, including its type as an explicit slot, with a `"data"` slot for which a JSON list is adequate and a slot `"missing"` to explicitly denote elements that should be `NA` in R. For details and an example, see page 300.

## Arbitrary server objects

XR also has a mechanism for representing arbitrary objects from the server language. Although the mechanism is general, it is used most often to represent objects from a server language class defined by a set of named fields/slots.

The server language code responsible for returning a value to R will use the ".RClass" mechanism to return an R object of class "from_Server" that represents the particular server language object. This class has slots for the server class, the module and the language, plus slots to accomodate essentially arbitrary structure. Specifically, the "data" slot holds any R object while the "fields" slot is expected to be a list with names corresponding to the named fields of the server language object. The contents of the relevant slots will be set by the server language computation that responds to the request for a converted version of the object.

In the case that the server class is defined by its fields, the converted object can be used in much the same way as the proxy class objects in the previous section. The `$` operator will extract or set fields and the valid fields will be those in the server language class. The difference is that the computations are now all being done in R and the contents of the fields will have been converted, by whatever method applies to *their* class.

Particular server language interfaces may subclass "from_Server" if that simplifies handling some special server classes. The XRJulia package defines a class "from_Julia". In this case, actually, most of the specialization is done on the Julia side, taking advantage of Julia's functional OOP. For some examples of general conversion, see Section 15.6, page 335.

The application package can request an arbitrary result to be returned by supplying .get = TRUE to an evaluator method or a proxy function. The server language side of the interface will convert basic objects, usually through the JSON mechanism, and will recognize other classes of objects for which specific R analogues exist (for example, arrays in Julia). Then, generally, an object from a server class defined by its fields will be converted by the "from_Server" mechanism provided that the objects in the fields are themselves convertible. This is a general, recursive procedure for converting arbitrary objects.

The limitation is that some "basic" objects may simply have no corresponding R object. Either the specific interface or the application will then need to come up with some combination of:

- server-side computations that recode the information in a form that does correspond to an R object; and

- R computations that interpret this converted object in terms of the information in the server object, as that applies to the particular application.

The XR structure cannot solve the specific problems, but it allows arbitrary conversions in the sense that if there *is* a solution, then the conversion should fit into the "from_Server" class. The mechanism is the "data" slot in that class, which is essentially for "anything else".

As an example, suppose that some class or data type in the server has no R analogue, and the decision is to serialize the server object. If the serialized object can be kept as a "raw" vector in R, then the object can be converted to a "from_server" object with that vector in the "data" slot, no fields and the appropriate "serverClass" and "module" values.

We may not seem to be much further ahead but there are two key advantages. Most importantly, this allows conversion of objects that have fields of a non-convertible type, making the rest of the object available. Also, the R object can be passed back to the server side of the interface, presumably without loss of information.

### Sending an R object

The character string to be passed to the server is defined to be the value returned by the generic function:

```
asServerObject(object, prototype)
```

The function will be called from an evaluator whenever an R object needs to be passed to the server language, typically as an argument to a function or method call. You can see the expression explicitly by calling the corresponding evaluator method:

```
ev$AsServerObject(object)
```

In fact, all the conversions to a server language expression in the interface methods are generated by a call to this method.

In calls to asServerObject(), the object argument is the R object to convert. The prototype argument represents the target for the conversion. In the evaluation methods, this object is the $prototypeObject field of the evaluator and has a class specifying the interface; for example, "PythonObject" or "JuliaObject".

The JSON-compatible text string for conversion to the server is returned by a call to another function, objectAsJSON(). The default asServerObject() method shows how this works:

```
function(object, prototype) {
    jsonString <- objectAsJSON(object, prototype)
    if(is(jsonString, "JSONScalar"))
        jsonString
    else
        gettextf("objectFromJSON(%s)",
                typeToJSON(jsonString, prototype))
}
```

The translated string for a scalar has a class that inherits from `"JSONScalar"`. The method assumes that the scalar format will be legal as a constant expression in the server language. For all other objects, the method constructs a call to the server language function `objectFromJSON()`, with the JSON string as the argument. Each server language interface will have a definition for this function, usually a simple call to the language's module for dealing with JSON.

The non-default methods for `asServerObject()` in the XR package all return names for objects in the server language. Methods for proxy and proxy class objects return the character string containing the unique server language key generated by the interface. The XR strategy assumes that proxy objects have been assigned in some environment in the server language using the key as the name. The character string key is then a reference to the corresponding server object.

The method for R objects of class `"name"` returns that name as an unquoted string. If an application makes an explicit assignment:

```
ev$Command("piBy2 = %s", pi/2)
```

then `as("piBy2", "name")` as an argument to `asServerObject()` will refer to the assigned object. Generally, however, it's better to use the unique keys generated by the interface to assign objects automatically.

The value returned by `objectAsJSON()` in the method shown above is the JSON string representing the R object. The string is substituted into a server language call to `objectFromJSON()`, unless the R object is to be treated as a scalar. So, for example, the R object `1:4` would be sent to the server as the string

```
objectFromJSON("[1,2,3,4]")
```

The function `objectAsJSON()` is another generic function. The JSON string returned by `objectAsJSON()` must represent a scalar, a list or a dictionary. Methods for R environments and named lists produce a dictionary directly.

R itself has no scalars, but `objectAsJSON()` interprets vectors of length 1 as scalars in the server language for the types that JSON can represent. This will likely be the right strategy most of the time, but to suppress scalars (e.g., because the server language function requires an array for this argument), pass the object to the interface through the call:

```
noScalar(object)
```

The `noScalar()` function uses an internal convention to force the string produced to be a JSON list.

Other objects will use the default `objectAsJson()` method, unless a particular interface package or an application has added more methods. The default method distinguishes two cases: basic data types in R and all objects with more structure, including formal classes, informal S3 classes and vector structures such as matrix and array.

Here are some examples to illustrate:

```
> prototype <- ev$prototypeObject
> objectAsJSON(1:3, prototype)
[1] "[1,2,3]"
> objectAsJSON(list(1,2,3), prototype)
[1] "[ 1.0, 2.0, 3.0 ]"
> objectAsJSON(list(a=1,b=2,c=3), prototype)
[1] "{ \"a\" : 1.0, \"b\" : 2.0, \"c\" : 3.0 }"
> objectAsJSON(1, prototype)
An object of class "JSONScalar"
[1] "1.0"
> objectAsJSON(noScalar(1), prototype)
[1] "[ 1.0 ]"
```

## Basic data types in **JSON**

Basic data types in R include vectors of various flavors, special types for representing the language and functions, class "NULL", and some specialized types such as external references, byte-code and others.

The conversion to JSON creates a list for those vector types that map individual values into JSON scalars: "numeric", "integer", "logical" and "character", with some special consideration for missing values. Essentially everything else is sent in the form of an explicitly identified R class. The representation of the class tries to retain all the information in the original data, encoded in some suitable form. Whether and how this information can be used is then up to the server language software.

Numeric data in JSON includes floating-point and integer-like notation. The format does not explicitly differentiate integers from numeric (float). The conversion in XR extends JSON notation so that, for example, the integer R value 1L is translated as "1" but the numeric 1 is "1.0", where JSON makes no distinction. All values in "numeric" vectors for conversion are adjusted by appending ".0" to integral values:

```
> objectAsJSON(1:4) # integer
[1] "[1,2,3,4]"
> objectAsJSON(1:4+0.) # numeric
[1] "[1.0,2.0,3.0,4.0]"
```

For floating point numbers, JSON does not include features of the floating-point standard for not-a-number (NaN) and for infinite values (Inf and -Inf in R). What are referred to as the "JavaScript extensions" to the notation do provide these. The XR conversion includes these in its output; the JSON modules in different server languages may or may not cooperate (yes in Python; no in Julia).

A more general conversion problem is that R has the concept of a missing value, NA, for several basic types. In numeric data, this can usually be substituted by the standard's NaN, but most server languages will have no equivalent for other basic data types. In this case, XR follows jsonlite in substituting a character string "NA". That is likely to cause exceptions on the server side, but no general, acceptable solution exists.

Other types have no direct analogue in JSON, including the types for function definition, language objects and the specialized non-vector types, such as external pointer. Type NULL is sent as the approximate JSON equivalent "null"; class "name" (type "symbol") is transmitted as its string value.

Objects from the remaining types are either deparsed or serialized to obtain what should be an equivalent character form. The resulting data is then enclosed in the explicit R representation used for S4 classes; that is, dictionaries with special element ".RClass" (page 288). The resulting string is passed as an argument to objectFromJSON in the server language.

Objects without JSON equivalent also include vectors of types "raw" and "complex". These are formatted into character vectors and once again sent in the explicit representation as dictionaries with ".RClass".

As an example, and to illustrate one useful way of studying the process, let's create a little vector of type "complex".

```
> z <- complex(real = c(1.5, 2.5), imag = c(-1.,1.))
```

To see the conversion process, we can call objectAsJSON() and then parse the result, as the server language would, but using the fromJSON() function in package jsonlite:

```
> zJSON <- objectAsJSON(z)
> jsonlite::fromJSON(zJSON)
$.RClass
[1] "complex"

$.type
[1] "complex"

$.package
```

```
[1] "methods"

$.extends
[1] "complex" "vector"

$.Data
[1] "1.5-1i" "2.5+1i"
```

While `"complex"` vectors are not formal classes, this representation is what they
would look like if they were: `".Data"` is the pseudo-slot that can always be used
to refer to the data part of a vector structure.

Having an explicit dictionary representation of an object from an R class allows
the server side of the interface code to include methods that interpret the R object
in whatever form is suitable.

## Methods for sending objects

The conversions are customized by methods, either for `asServerObject()` or for
`objectAsJSON()` if the method only needs to alter the standard representation
as a JSON string. Methods in the XR package for these functions are language-
independent, with only the `object` argument in the signature. Methods in interface
packages are likely to be specialized to that language. For example, conversions
are specialized for sending `"array"` objects to Julia by:

```
setMethod("asServerObject", c("array", "JuliaObject"),
        function(object, prototype) {
            data <- asServerObject(as.vector(object), prototype)
            dims <- paste(dim(object), collapse = ",")
            value <- gettextf("reshape(%s, %s)", data, dims)
            value
        })
```

This uses the Julia function `reshape()` to convert a one-way array to the appropriate multi-way form.

The method recalls `asServerObject()` to convert the `"vector"` of data and
inserts the text for this conversion into the expression being constructed. The
conversion of a vector to Julia produces a one-way array, suitable as the argument
to `reshape()`.

Another detail in this example is worth noting. Julia requires each element
of `dim()` to be a separate argument to `reshape()`. The call to `paste()` in R
constructs this variable length call in Julia. The `"dim"` slot does not turn into

a field in Julia but into part of the expression. Constructing the server language expression as a string allows this sort of flexibility.

The general conversion technique based on JSON is convenient and facilitates our goal of supporting arbitrary computations. But it may not be the best choice when the data involved are large or have a structure quite different from typical R objects. One may then prefer to use some special-purpose conversion mechanism available both in R and in the server language.

XML data, as used in the shakespeare examples, illustrates the need for special treatment. XML is itself a standard for data representation. All the usual server languages, and R itself, have modules to deal with such data and one or more internal representations for it. The obvious conversion mechanism is to write and read XML itself as the intermediate form. Encoding an XML object via the general form would be difficult to define, inefficient and potentially inaccurate for complex examples.

In the example on page 280, a proxy for the parse() function in the Python module converted XML on a file into an "ElementTree" object in Python. The R package XML has several classes for representing such data. If an application is producing XML objects in R, it might like to send the objects to Python. This would be implemented by a method for asServerObject(). The object to be converted would be from a suitable class that can be saved to a file, such as "XMLInternalDocument" in the XML package. The prototype class in the signature can be left as a general "PythonObject" if we want to use this conversion by default when sending XML to Python.

A simple version of the method creates a temporary file, saves the R object to it, and constructs a Python expression to parse the file:

```
setMethod("asServerObject",
    c("XMLInternalDocument", "PythonObject"),
        function(object, prototype) {
            file <- tempfile()
            XML::saveXML(object, file)
            gettextf("xml.etree.ElmentTree.parse(%s)",
                    asServerObject(file, prototype))
        })
```

asServerObject() is recalled to convert the string in file to a string in the server language. In practice, a somewhat better method would be needed.

The R object was written to a temporary file. This file has to stay around until the corresponding Python method reads it, but then it should be removed. Rather than the parse function in the "ElementTree" module, we should write a specialized Python function, say XMLFromR, that ends by removing the file.

There is an important limitation to `asServerObject()` methods that replace the JSON-based set of methods for conversion. The approach is fine to convert an object from a given class. But if the computation is converting an object that *contains* such an object, it is the method for the containing object that counts. If that method constructs a JSON string, then our method for `asServerObject()` will not be called; instead, the conversion is likely to call `objectAsJSON()` with the XML data as argument. That will happen if our XML object is an element in an ordinary list or is a slot in an object that will be converted using the general representation discussed above.

What to do? A reasonable approach might be something like this:

- Specialized methods replacing the standard approach are mainly useful for seriously large and/or complex objects. These should usually be converted once and then utilized in the server language. The strategy would then be to construct containers and higher-level objects in that language.

- Where it does make sense to convert higher-level objects all at once, give these their own class, with an appropriate `asServerObject()` method for conversion. A list of XML objects could have a class `"listOfXML"` with an `asServerObject()` method.

- If the decision is that the JSON-based approach is unacceptable, rather than just slightly sub-optimal, include a method for `objectAsJSON()` that signals an error.

The method for a higher-level object containing one or more of our specially converted classes will often work best through a custom function in the server language. For example, suppose the server language had a function `listOf()` taking an arbitrary number of arguments and making a list-like object with each of the arguments as an element. Then the essential computation in a `"listOfXML"` method would be

```
calls <- sapply(object, function(x) asServerObject(x, prototype))
paste0("listOf(", paste(calls, collapse = ", "), ")")
```

The first line creates a character vector with the individual expressions to convert the elements, such as calls to our `XMLFromR()` function, the second line constructs a single, potentially large call to `listOf()`.

See the treatment of multi-way arrays in Julia, in Section 15.6, for another example.

### Getting a server object

The value returned from the server is always a string. This is first parsed using JSON notation.

Since JSON has no typed arrays, an R vector of a basic type sent and then returned using JSON notation will become a list. The user (or the specializing server language interface package) can avoid this by setting the field "simplify" in the evaluator to TRUE. Any list object in JSON whose elements are all basic scalars will then be turned into an R vector of the necessary type. The XR package defaults to simplify = FALSE so as not to interfere with specializing strategies.

The R object obtained from parsing the JSON string is then passed as the object argument to the generic function:

```
asRObject(object, evaluator)
```

The value of the call will be the actual object returned from the server. The default method just returns object, but custom methods can do anything that suits the application. The evaluator argument allows the method signature to specify computations for a specific interface class. Unlike an initialize() method, there is no requirement that asRObject() returns any particular class of object, or that it always returns the same class.

The direct conversion procedure can produce an object from an arbitrary R class, using the explicit dictionary representation illustrated starting on page 288. To do so, the function or method in the server language needs to construct a dictionary object, or anything that will be returned to R as such. The object must have a ".RClass" element with the appropriate class name and other elements with the names of the slots in the R class definition.

After the conversion produces object from a class through the special dictionary representation, XR interface computations will call asRObject() for this object.

For the computation to produce a valid object in the target R class, each of the slots must be converted to an object from the class specified for that slot in the R class definition. If the target class for some slots is inconvenient to generate from the server language, it may be best to define an intermediate R class with slots that *are* convenient and that contain the essential information to define the target class or classes via computations in R. The mechanism to return typed R vectors is an example.

### Class "vector_R"

There is considerable variation among server languages in how they treat objects analogous to "vector" classes in R. Some languages have as basic data types only something similar to `"list"` in R: array-like objects with arbitrary data in each element. This is also the case for the basic JSON notation underlying the direct conversion procedure in XR.

Conversion of a general list to a chosen vector type is provided in the XR package by the class `"vector_R"` with slots `"type"` and `"data"`:

```
setClass("vector_R",
         slots = c(data = "vector", type = "character",
                   missing = "vector"))
```

The important distinction is that the object returned from the server side can have a `"data"` slot that is a list.

To return a vector of a chosen type, the server side must provide a suitable mechanism to generate an object from the `"vector_R"` class. XRPython and XRJulia have a corresponding server-language function `vector_R()`. Like JSON, Python does not have typed vectors, so an application package will need to be explicit when wanting to have the R object be a vector of a specific type, by calling `vector_R()`. But all the arguments can be standard Python objects—lists and character strings—so no exotic computations are needed.

Julia does have typed single- and multi-dimensional arrays. Conversion to R via the `vector_R()` class is automatic, with no action on the part of the application needed. To implement the return of suitable `"vector_R"` objects, the XRJulia interface uses Julia's own generic functions and functional OOP. In effect, `vector_R()` implements the method for one-way arrays of the generic function `toR()` in XRJulia, which transforms an arbitrary Julia object into the form needed for transmission to R. See Chapter 15 for the details.

In addition to providing type information, the class `"vector_R"` provides explicit information about missing values. Neither JSON nor most server languages have a uniform mechanism for indicating missing values from vectors of different types. Standard floating point representation has a somewhat analogous concept of `NaN` for not-a-number, but integer, logical or character string data has no similar mechanism.

Objects from `"vector_R"` have a slot `"missing"`, which is interpreted as a vector of indices for all elements of the vector that should be treated as `NA`. As with the type information, missing value information can be generated automatically in R and passed to the server language or can be explicitly computed and returned to R by constructing the R representation, including the `"missing"` slot.

## Objects for exceptions

Another important application for returning an explicit form of an R class comes when handling errors and other exceptions. The XR design requires that an inter-

face package for a particular language return an object from a subclass of class
`"InterfaceCondition"` when an error or other exception arises. The subclasses
include `"InterfaceError"` and `"InterfaceWarning"`, expected to be treated as
errors and warnings in R. There can also be any specialized subclasses that make
sense for a particular language or application. All the classes will, as always, inherit
the slots of their superclass, `"InterfaceCondition"`:

```
setClass("InterfaceCondition",
    slots = c(message = "character", value = "ANY",
              expr = "character", evaluator = "Interface"))
```

The `"message"` slot is the descriptive message for the error or other exception, as
provided by the server language (one hopes). The `"expr"` slot is the expression
provided by the R side for evaluation. These are both character strings.

The `"value"` slot is the value of the expression that would have been returned
in the absence of the exception. For an error, `value` is ignored. In the case of a
warning exception, it will normally be returned after the user is notified of the
exception. From a data conversion perspective, the `value` slot requires nothing
special, but is handled exactly as the result of the computation would have been
without the exception.

The `"evaluator"` field has the evaluator object being used when the exception occurred, and is inserted by the R side of the interface. The XR function
`doCondition()` is expected to be called from the `$ServerEval()` method of a
particular interface. This is another generic function. Its default mechanism will
signal the corresponding R condition, with information about the interface added
to the message in the object. Applications can further specialize condition handling
by defining new classes that extend `"InterfaceCondition"`.

Methods for `doCondition(object)` can specialize the handling. To inherit the
actual condition behavior, the method should end with

```
callNextMethod()
```

In particular, this will eventually call `stop()` or `warning()` for conditions that
extend `"InterfaceError"` or `"InterfaceWarning"`.

Before this point, `asRObject()` will be called with the `"InterfaceCondition"`
object as an argument. If the application wanted to do something completely
unrelated to R condition handling, it could define a method for `asRObject()` that
returned a different object, perhaps conditional on the circumstances.

# Chapter 14

# An Interface to Python

## 14.1   R and **Python**

The XRPython package implements an interface to Python according to the XR interface structure described in Chapter 13. This chapter describes the use of the interface in an application extending R. The application will likely implement an R package using the interface to carry out some computations in Python and incorporating those computations into software for the users of the application. The XRPython package, the XR package it imports and the shakespeare package examples are available from `github.com/johnmchambers`.

Python computations consist mainly of function calls and the use of methods and fields in objects from Python classes. These are available directly through methods for an interface evaluator or equivalent function calls (Section 14.2).

Extending R using the interface to Python usually includes providing functions in an application package in R. These can incorporate such direct computations in Python.

Additional methods and function calls provide interfaces to other Python programming that is likely to be needed, such as importing modules (Section 14.3)

If specific functions or classes in Python are useful, *proxy* functions and classes in R (Sections 14.4 and 14.5) make a much simpler interface. Calls to the function and computations on the objects appear to the user as ordinary R expressions, but carry out the corresponding computation in Python.

The XR interface structure creates and manages proxy objects for the results of computations in Python, avoiding conversion of data between the languages when not required. Data conversion facilities are available when needed, including the representation of arbitrary R objects in Python and the return of arbitrary Python objects to R (Section 14.6).

Python is a valuable resource for interface programming. It is a popular language for implementing a wide range of computational techniques. Its design emphasizes readability and simplicity. From our perspective, it has many modules covering all sorts of techniques. Programming in Python, particularly to extend existing software, is convenient and much of the programming structure meshes smoothly with R and with the XR structure for interfaces.

Applications using the interface typically will do some Python programming of their own to supplement or specialize existing software. The R package structure facilitates this by having a directory of Python code installed in the R library. The code will typically consist of one or more Python modules that are imported by the package. When changing and testing the modules, it may be more convenient to work directly in Python, through an interactive environment such as supplied by the Jupyter application[1].

The XRPython package is designed for computing with R, with help from an interface to Python as a server language. Projects starting from Python could do the opposite. There is a widely used interface *from* Python to R, rpy2.[2] If you're working from a Python project, take a look at that. However, our target is to contribute to a project with some substantial programming and interaction in R; programming in the large, in the sense of Chapter 8. The INTERFACE principle encourages such projects to reach out to helpful computations, in this case written in Python.

## 14.2   Python Computations

Computations in Python using the XRPython interface are carried out by a Python interface evaluator, an object from class `"PythonInterface"`. The current Python evaluator is obtained by:

```
ev <- RPython()
```

If there is no such evaluator, one will be created, starting the embedded version of Python. Although it is possible to have multiple active evaluator objects, there is usually no reason to do so with an embedded interface.

All the interface computations use methods and fields of an evaluator object. R programmers may feel more comfortable with a functional computation rather than methods. All the methods have functional equivalents as we will describe below. There is no operational difference for the computations in this section, but some differences for the programming operations of Section 14.3.

---

[1]http://jupyter.org
[2]`rpy.sourceforge.net`

The computations in this section are the base for all other computations such as those using proxy functions and classes. They provide maximum flexibility for implementing interface calculations, but the proxy techniques are simpler when they apply. For the user of the application package, in particular, computations are expected to appear as regular R with no need to refer explicitly to the interface. An application package can hide the interface through any combination of functions in R using direct interface computations and proxies to software in Python incorporated into the package.

The main interface methods for direct computation are `$Eval()` for expressions returning a value and `$Command()` for other Python statements and directives:

```
ev$Eval(expr, ...)
ev$Command(expr, ...)
```

with functional equivalents:

```
pythonEval(expr, ...)
pythonCommand(expr, ...)
```

These and the other functional equivalents for computations obtain the current evaluator object and call the corresponding method.

In these computations `expr` is a character string to be parsed and evaluated in Python. Remaining arguments are objects, either proxy objects from previous expressions or R objects to be converted to Python. The objects are inserted into `expr` to replace C-style `"%s"` fields in the string.

All expressions evaluated by the interface are computed by Python function `value_for_R()`. If the type of the result is one of the standard scalar types (numeric, logical, string or `None`), the value is converted to a string that R will evaluate to get the equivalent R object. Otherwise, the result is assigned in Python. The string returned will be interpreted in R as a proxy object containing the name used for the assignment and a description of the result (the Python class, the length and optional module). The proxy object will turn into its name when used in a subsequent calculation. Thus the result of any interface computation can always be passed back to Python for further use.

To force a computed result to be converted, the `$Eval()` method and all proxy functions take an optional argument `.get=`; supply `.get = TRUE` to force conversion.

Objects may also be converted explicitly. The `$Send()` method converts its argument to Python and returns a proxy object for the result. The `$Get()` method converts a proxy object to the equivalent R object, as far as possible. Both methods will do something for arbitrary objects; see Section 14.6.

To illustrate these methods, first we create a Python list of 3 numbers:

```
> xx <- ev$Eval("[1, %s, 5]", pi)
> xx
R Object of class "list_Python", for Python proxy object
Server Class: list; size: 3
```

The value returned is a proxy for the list. The first line of the example could equally have been:

```
xx <- ev$Send(c(1, pi, 5))
```

We can print any result in Python:

```
> ev$Command("print %s", xx)
[1, 3.14159265358979, 5]
```

In Python, unlike in R, printing is not an expression.

Python lists are one of the built-in proxy classes, as the first example showed. The proxy object xx has all the usual list methods:

```
> xx$append(4.5)
NULL
```

To get the Python list as an R object:

```
> as.numeric(ev$Get(xx))
[1] 1.000000 3.141593 5.000000 4.500000
```

Python has no typed arrays, so we convert the list of numbers to a numeric vector on the R side. An alternative is to construct a special vector object in Python and convert that:

```
> ev$Call("vectorR",xx, "numeric",.get = TRUE)
[1] 1.000000 3.141593 5.000000 4.500000
```

(See Section 14.6, page 319, for the technique and another example).

Objects in the interface evaluator are valid only while that evaluator is running; similarly, R proxy objects will no longer be valid afterwards. To preserve proxy objects over sessions, serialize them to a file:

```
> ev$Serialize(xx, "./xxPython.txt")
```

Then, at some later time, the $Unserialize() method will bring the object back, to a new evaluator:

```
> xxx <- ev$Unserialize("./xxPython.txt")
> xxx
Python proxy object
Server Class: list; size: 4
> as.numeric(ev$Get(xxx))
[1] 1.000000 3.141593 5.000000 4.500000
```

The serialization is done Python-style, with `pickle()`, so issues of conversion do not arise. The methods have optional arguments as described in Section 13.6, page 278.

All these have functional equivalents with the same arguments as the methods:

```
pythonGet(object); pythonSend(object)
pythonSerialize(object, file); pythonUnserialize(file)
```

# 14.3 Python Programming

In addition to the general methods in the previous section and to the proxy objects, functions and classes, the XRPython interface provides some methods to assist in using Python software and in extending the software. The software available can be extended by adding to the search path for Python modules and importing modules. Python source can be included directly from source. An interactive Python shell can access the objects computed in the interface.

Python's modules are files of source, possibly compiled. The Python evaluator looks for these in a system path list of directories, similar to R's `.libPaths()`. XRPython has two functions for appending a directory to the system path and for importing from modules found on the path.

To add a directory to the system path in Python:

```
pythonAddToPath()
```

The expected use of the call is to make a directory of Python code in an application package available. Application packages will usually need to write some of their own functions, classes and other computations in Python. The XR structure for interfaces has a convention that the Python code in an application is in the `"python"` subdirectory of the installed package,

The call with no arguments, as above, from the R source in an application package adds the directory named `"python"` in the installed version of the package from which the call originates.

If the application wants to use a different source directory, perhaps in order to have different versions or optional additional code, supply the directory as an argument; for example, `"pythonA"` for source directory `"inst/pythonA"`.

The `package=` argument should be given explicitly if the code to be added is from a different R package:

```
pythonAddToPath("plays", package = "shakespeare")
```

This would add directory `"plays"` in the R package shakespeare, which would correspond to directory `"inst/plays"` in the source for that package. To add a directory that is not part of any R package, give an empty string, `""`, as the `package` argument. For more details on the XR structure for path management see Section 13.4, page 270.

Once the necessary directories are on the system search path, Python has a variety of commands to import modules (files) and make their contents available. These are expressed with an R-style syntax through `pythonImport()`. The first argument in the call is a module name, the remaining arguments are the names of objects to be explicitly imported from the module. If `"getPlays.py"` is a file of Python code in a directory that has been added to the path, then

```
pythonImport("getPlays", "byTitle")
```

will import the object `byTitle` created by the code in that file. For those familiar with Python, this is in the style of the Python command

```
from module import ...
```

All the arguments should be character strings, but the `"..."` arguments can be vectors with multiple object names. The evaluator keeps track of imports, so repeated calls for the same imports only go to Python once. The call can be used to import any objects, not just functions.

Giving only the module argument is equivalent to the Python `import` directive. This is different from a simple `import` call in an R namespace, in that it does not make the simple object names available. So

```
pythonImport("getPlays")
```

would not make `"byTitle"` available by simple name, but would recognize the fully qualified version, `"getPlays.byTitle"`. The Python convention to import all objects is invoked by providing the second argument as `"*"`.

If functions are handled by defining proxy functions in R as described in Section 14.4, there is no need to import them or their module explicitly; the proxy function takes care of that.

The functions `pythonAddToPath()` and `pythonImport()` have corresponding evaluator methods. These differ in that they only affect the particular evaluator on which they are invoked. The function versions will apply to any evaluator, and therefore can be called before an evaluator exists. An application package can call them from its source code: the actions will take place when the package is loaded. For any likely situation, the function version is more appropriate.

The remaining programming methods to be described can be called equivalently as method or function, with identical results.

To evaluate a file of Python source, use `pythonSource()` or the method:

```
ev$Source(filename)
```

The file can contain arbitrary expressions and commands. They will be evaluated in the same namespace as other evaluator expressions, so assignments will become available for later computations. To refer to an explicitly assigned object as an argument in an interface expression, the name should be supplied as an R object of class `"name"`, *not* as a character string. So, if our source file had the line

```
pi = 3.14159
```

then the object assigned as `pi` should be referred to by `as.name("pi")`:

```
> ev$Eval("%s/2", as.name("pi"))
[1] 1.570795
```

Proxy objects are nearly always a simpler form of reference, however.

For the definition of a single function XRPython also has a `$Define()` method which can take as an argument text containing a Python function definition. It sends the function definition to Python as a command and returns a proxy for the newly defined function. For a trivial example, consider the function definition:

```
def repx(x):
    return [x, x]
```

For the `$Define()` method this is equivalent to a character vector with the lines of the definition as elements:

```
> text <- c("def repx(x):", "    return [x, x]")
> repxP <- ev$Define(text)
```

This creates the R proxy function, assigned as `repxP`:

```
> twice <- repxP(1:3)
> unlist(pythonGet(twice))
[1] 1 2 3 1 2 3
```

The definition can also be taken from a file, via the `file=` argument.

```
>repxP <- ev$Define(file = "./repx.py")
```

The effect is to read all the lines from the file and collapse them into a string,
separated by new lines. If one does not need to create the proxy function in the
same call, $Source() is more efficient.

To do some interactive computations in the same workspace used by the evaluator, through the $Shell() method or the function version:

```
pythonShell()
```

This runs a simple interaction, reading expressions and evaluating them through
the command method. To see results, you need to use the Python print directive,
or call some other output-generating function. To exit the evaluator, give the usual
Python command, "exit".

```
> pythonShell()
Py>: print repx([1,2,3])
[[1, 2, 3], [1, 2, 3]]
Py>: exit
```

The interactive environment is definitely not competitive for Python programming;
in particular, to continue an expression over multiple lines, you need to end all
the lines except the last with an unescaped backslash. The advantage of $Shell()
is that expressions use the same working environment as other interface methods.
Another way to define a Python function directly would be to enter the definition
to the shell, escaping all but the last line (and being careful to follow Python's use
of spaces to define blocks of code).

```
> pythonShell()
Py>: def rep3(x):\
Py+:    return [x, x, x]
Py>: print rep3(pi)
Python error: Evaluation error in command "print rep3(pi)":
     name 'pi' is not defined
Py>: pi = 3.14159
Py>: print rep3(pi)
[3.14159, 3.14159, 3.14159]
Py>: exit
```

As the example shows, errors in Python are reported but the shell continues until
a line matches "exit". The interactive expressions are evaluated in the same
context as other interface method calls, meaning that you can query various system

parameters, such as the search path, to better understand what is happening to your interface computations.

In order to examine proxy objects in Python, you need to get the name under which the Python object was assigned, by calling pythonName() (outside the shell):

```
> pythonName(xx)
[1] "R_1_1"
> pythonShell()
Py>: print R_1_1
[1, 3.14159265358979, 5, 4.5]
Py>: exit
```

# 14.4 Python Functions

Functions are central to Python; it operates largely following its own version of the FUNCTION principle. Low-level computations such as arithmetic are not function calls, at least not when they apply to basic objects. Other than that, pretty much everything is, including methods in Python's encapsulated OOP implementation. This is functional *computing* as opposed to functional *programming*, as distinguished in Section 1.5. Objects are references and side effects from function calls are to be expected.

Calling Python functions from R is made simpler by the use of proxy functions; that is, functions in R, calls to which map directly to corresponding Python function calls. For most functional computing, proxy functions eliminate the need to deal directly with the interface evaluator.

A proxy to a Python function is created by calling PythonFunction() in the XRPython package:

```
PythonFunction(name, module)
```

This returns a proxy function object given the name and, optionally, the module from which the function should be imported. A call to this function object in R evaluates a corresponding call to the named Python function, returning (usually) a proxy object for the result of that call. PythonFunction() is actually the generator function for the "PythonFunction" class, a subclass of "function" with extra information describing the Python function.

An example: The "xml" module in standard Python has a function parse() down a few levels, in module "xml.etree.ElementTree". To create a proxy function for parse():

```
> parseXML <- PythonFunction("parse", "xml.etree.ElementTree")
```

A call to the R function imports the Python module if necessary and calls the function:

```
> hamlet <- parseXML("./plays/hamlet.xml")
> hamlet
Python proxy object
Server Class: ElementTree; size: NA
```

The proxy object returned may be from any relevant Python class, in this example "ElementTree". The Python function can also be written specially to return the representation of a general R object as discussed in Section 13.8, page 288, in which case the interface will construct an object from the target R class.

For extending R, proxy functions should be defined in the source of an application package. Users of that package need not be concerned with the proxy nature of the function: parseXML() can be treated by users as a regular R function.

Proxy functions mesh with proxy classes. If the Python methods or fields for the "ElementTree" object are useful (as they are, in fact), the application package will define a proxy for the Python class. As with the proxy function, users of the application package can then invoke methods or access fields just as they would for an R reference class, without worrying about the interface. We'll look at this example in the next section.

Let's examine the actual R function generated in the example above:

```
> parseXML
Proxy for Python function "parse", from module "xml....
function (..., .ev = XRPython::RPython(), .get = NA)
{
    nPyArgs <- nargs() - (!missing(.ev))
    if (nPyArgs < 1)
        stop("Python function parse() requires at least....
            nPyArgs)
    if (nPyArgs > 2)
        stop("Python function parse() only allows 2 ....
            nPyArgs)
    .ev$Import("xml.etree.ElementTree", "parse")
    .ev$Call("parse", ..., .get = .get)
}
```

In addition to arguments that will be passed to the Python function, all proxy functions have two optional arguments: .ev= and .get=: the first of these specifies the evaluator to be used and the second the strategy for converting the result to an R object. The evaluator is by default, and nearly always, the current Python interface evaluator.

The default strategy for conversion corresponding to .get=NA is to convert scalars and return everything else as a proxy object. To force all results to be converted, include the argument .get = TRUE.

In constructing the proxy function, XRPython uses metadata in Python about the function object. In this example, the Python function has two arguments, one required and one optional. After checking for too few or too many arguments, the body of the proxy function does two things: it ensures that the Python module is imported in the form needed and it evaluates a call to the Python function using the $Call() evaluator method.

Python has several versions of importing, presenting options for proxy functions. The call to PythonFunction() above implied that the object named "parse" should be imported from the specified module and called as parse(). The $Import() method call in the proxy function does this, effectively generating the Python directive

```
from xml.etree.ElementTree import parse
```

The alternative is to import only the module and then to call a fully qualified reference to the function. This would be done in the example by giving the fully qualified reference to the function entirely in the function name:

```
> parse2 <- PythonFunction("xml.etree.ElementTree.parse")
```

The resulting function will generate similar Python calls, but to the fully qualified version of the function, and will generate the simple import statement:

```
import xml.etree.ElementTree
```

The two versions will do the same computations but only the first will define a local variable "parse" in the namespace. The distinction is similar to that in R between importing a function into a package and referring to it by an expression like shakespeare::parseXML. The tradeoff is similar too. Importing the object explicitly leads to more readable code but could mask a relevant object of the same name in some standard module, just as having parse() in the R application package could in some situations mask the parse() function in base.

The argument list of the proxy function is passed directly to Python, aside from checking the number of arguments. Python allows much of the flexibility of R in argument calls, such as named arguments ("keyword" in Python terminology), and missing arguments to some extent, provided the arguments have default values (precomputed, not expressions as in R). The XRPython proxy functions pass along

the pattern of positional and named arguments in the R call. The proxy calls the
Python function with the arguments supplied, including names.

In addition to regular arguments, required or optional, Python has a mechanism
(two, actually) for an arbitrary number of arguments, similar to "..." in R. A
formal argument in Python of the forms "*name" and "**name" match arbitrarily
many unnamed and named ("keyword") arguments. If either of these patterns is
present, the interface allows arbitrarily many arguments to the Python function.

The specific list of formal arguments is available from the Python metadata,
and is retained in the proxy function for documentation as slot "pyArgs":

```
> parseXML@pyArgs
[1] "source"   "parser ="
```

The trailing "=" indicates optional arguments. If the function took arbitrarily
many arguments, the last name in the list would be "...". Python functions can
have inline documentation, like reference methods in R. If they do, this is kept in
the R proxy as slot "pyDocs" (parseXML() doesn't have any).

# 14.5   Python Classes

Python implements encapsulated OOP by `class` definitions in the language, containing a set of method definitions. Evaluating the definition creates a corresponding class object. The class definition also creates automatically a generator function of the same name for instances of the class, defined by a special initialization
method. As in R, methods are slightly specialized Python functions stored as members of the class object.

Typically for Python, the implementation is simple to use and relatively simple
internally. There are very few restrictions, which makes for flexibility. But the
definition says nothing about fields, preventing validation of their types, or even
identifying the fields through the class definition. With a few such limitations, an
interface from R to classes in Python is straightforward and helpful.

A call to setPythonClass() in the XRPython package produces a proxy class
definition in R for a Python class, incorporating the Python metadata for the
class within the XR structure. When the call comes from the source code for
an application package, the metadata is needed during installation. To avoid this
requirement, a load action or a setup step may be used to define the class when the
application package is loaded or by a separate script (see Section 13.7, page 282).

Knowing the Python class and module is enough to obtain metadata that
defines the methods for the R class. The module must be available from the Python
search path when setPythonClass() is called, which may require an earlier call
to pythonAddToPath().

Because the fields are not formally defined, more effort may be required for them. If possible, an object from the class is used as an example. If the default object generated from the class has the necessary fields, this will be used automatically. Otherwise, an appropriate `example=` argument in the call will be needed, as shown below.

In the use of XML data in our shakespeare package, the object returned from the `parseXML()` is of Python class `"ElementTree"`, defined in the Python module `"xml.etree.ElementTree"`. A proxy R class is created by:

```
ElementTree <- setPythonClass("ElementTree",
                       module = "xml.etree.ElementTree")
```

The value returned by a call to `setPythonClass()` is a generator function for the class, as usual. It's more likely, however, that a suitable object is created by a call to a Python function, such as the example on page 312. If a proxy class has been defined for the server language class of the proxy object returned, the return value is automatically promoted to this class, making the methods and fields available.

In the example, the value returned is from Python class `"ElementTree"`, which has a method `findtext()`, among others. Once the proxy class is defined, the object returned can use all these methods directly:

```
> hamlet$findtext("TITLE")
[1] "The Tragedy of Hamlet, Prince of Denmark"
```

As it happens, this class has no interesting fields.

Because Python has no formal definition of the fields in a class (as with S3 classes in R), objects from the class will have fields with whatever name has been used in assigning them, usually via the `` `.` `` operator in Python. Fields must be inferred from an object or specified explicitly; there are some options in the call to `setPythonClass()` to handle different cases.

The assumption when only class and module are specified is that the default object from the Python class has the appropriate fields. The interface generates that object, by calling the Python generator function for the class with no arguments, and examines its fields. This works in the example since there are no useful fields anyway. It would also work if the class was specially designed for the R interface, as is the `"Speech"` class discussed later in the section.

More typical of Python programming is to generate a default object with few or no fields, and have various methods add the fields. In this case the application package should compute an object from the class that *does* have suitable fields, if possible, and supply a proxy for this object as the argument `example=` to `setPythonClass()`.

The elements of the parse tree in our example have **Python** class `"Element"`. A suitable object from this class is returned by the `getroot()` method for class `"ElementTree"`. A computation to define the `"Element"` proxy class could be:

```
> hamlet <- parse("./plays/hamlet.xml")
> Element <- setPythonClass("Element",
+                   module = "xml.etree.ElementTree",
+                   example = hamlet$getroot())
```

The output of

```
Element$fields()
```

will show the fields; a few of these are special fields for the interface; the rest, with class `"activeBindingFunction"`, are proxies for the **Python** field of the same name. In this case, the object has proxy field `"tag"`, among others. Once the proxy class for `"Element"` has been defined the fields are available:

```
> hamlet$getroot()$tag
[1] "PLAY"
```

Applications will often find it convenient to add their own **Python** code, including class definitions. These can be designed for use by proxy and be valuable in structuring the data conveniently for an interface from R.

In this example, the **XML** structure is very hierarchical, suitable for tree-walking logic and recursive computations. In the Shakespeare data, plays are organized into subtrees by `ACT` and these in turn into subtrees by `SCENE`. Within a scene there will be multiple speeches, a third level of subtree, along with other nodes for stage directions, etc.

Both R and **Python** tend to prefer a more linearized organization; in R to be natural for vectorized computation, in **Python** to simplify iteration. The **Python** software in **shakespeare** includes functions and corresponding classes to "flatten" the structure. In particular, one would often like to analyze the contents of individual speeches; for example, to compare speakers within a play or across plays.

Package **shakespeare** defines proxy **Python** classes for each level of the hierarchy and corresponding methods that flatten the tree into the form of a list of all acts, all scenes or all speeches. In particular, the **Python** function `getSpeeches()` returns a list of all the speeches in the tree supplied as its argument. Each element is an object of class `"Speech"`.

All the objects of this class have fields `"play"`, `"act"`, `"scene"`, `"speaker"` and `"lines"`. The first four fields are character strings. The last is a list of the lines of text in speech, extracted from the corresponding **XML** nodes.

The idea of the class is that objects are normally created from corresponding XML nodes of tag SPEECH. The initialization method expects to get such an object as its argument. It's a good rule, however, to make initialization methods (whether in Python or R) work sensibly when no argument is given. In this case, the class was designed to work with the R interface, so its initialization method in Python deliberately creates all the relevant fields:

```python
def __init__(self, obj = None,
             act = '<Unspecified>', scene = '<Unspecified>'):
    self.act = act
    self.scene = scene
    ## to be a well-behaved class, we always set the 4 fields
    if obj is None:
        self.speaker = '<Unspecified>'
        self.lines = [ ]
    else:
        self.speaker = obj.findtext('SPEAKER')
        lines = obj.findall('.//LINE')
        linetext = []
        for line in lines:
            linetext.append(line.text)
        self.lines = linetext
```

To set up some analysis of speeches, the user calls the R proxy for the function getSpeeches() with the XML tree previously parsed from the file for the play:

```r
> speeches <- getSpeeches(hamlet)
> speeches
R Object of class "list_Python", for Python proxy object
Server Class: list; size: 1138
```

The speeches object is a proxy for a standard Python list, whose elements are objects of class "Speech". Proxy classes for Python lists and dictionaries are built into the XRPython interface, so the methods for them can be used in R as they would be in Python. The shakespeare package has created a proxy class for "Speech".

For example, if we're curious about the last speech in the play:

```r
> last <- speeches$pop() # the last speech
> speeches$append(last) # put it back
NULL
> last
```

```
R Object of class "Speech_Python", for Python proxy object
Server Class: Speech; size: NA
> last$speaker
[1] "PRINCE FORTINBRAS"
> last$lines
R Object of class "list_Python", for Python proxy object
Server Class: list; size: 9
```

We'll come back to this example in discussing data conversion, the next topic.

# 14.6   Data Conversion

Python objects come in a range of built-in and user-defined classes (or "types", both terms being used in Python). For applications, as opposed to programming, the important built-in types include a variety of scalars: numeric of several flavors, character string and boolean. Other than scalars, the built-in container types are "list" for (unnamed) list objects and "dict" for dictionaries.

The default XR strategy sends named lists and environments to Python as "dict" objects; all other R vectors (of any type) are sent as "list", except that by default vectors of length 1 are sent as scalars. Some examples, in which the "Server Class" of the proxy object shows what was sent:

```
> pythonSend(1:3)
R Object of class "list_Python", for Python proxy object
Server Class: list; size: 3
> pythonSend(1)
Python proxy object
Server Class: float; size: NA
> pythonSend(1L)
Python proxy object
Server Class: int; size: NA
> pythonSend(list(first = 1, second = 2))
R Object of class "dict_Python", for Python proxy object
Server Class: dict; size: 2
```

As the first and last examples show, lists and dictionaries have proxy classes in R. All their usual Python methods should be available.

```
> a <- pythonSend(1:3)
> a$reverse()
NULL
```

```
> as.integer(pythonGet(a))
[1] 3 2 1
> b <- pythonSend(list(first = 1, second = 2))
> b$has_key("first")
[1] TRUE
```

We don't create proxy classes for the scalar types; they have few useful methods and will normally not be retained as proxy objects.

Vectors of length 1 by default will be sent as scalars. The `noScalar()` function turns off the scalar option (the contents of the object are unchanged otherwise):

```
> pythonSend("testing")
Python proxy object
Server Class: str; size: 7
> pythonSend(noScalar("testing"))
R Object of class "list_Python", for Python proxy object
Server Class: list; size: 1
```

Python has no typed arrays, unlike R or Julia. When an application needs to indicate the type of a vector in R, either when the object is being sent to Python or is being returned in the value of an expression, the technique is to use the R class, `"vector_R"` (Section 13.8, page 300). This class has a slot `"data"` containing a vector, but the type of the vector is explicitly provided as slot `"type"`, so that the intended type of the vector remains clear when the object is translated in JSON and also when it is sent to a language such as Python that will treat all basic R vectors as list objects.

When an object being converted is explicitly of class `"vector_R"`, a method for `asRObject()` will coerce it to the declared vector type. This can be used in Python functions to customize list objects when returned to R. The Python side of the XRPython interface has a function `vectorR()` with arguments

```
vectorR(obj, type, missing)
```

The last two arguments are optional, specifying the R class to which the converted vector should belong and a list of elements that should be interpreted as missing.

In our previous example with Python class `"Speech"`, the list in field `"lines"` is known to have only character strings as elements. The class has a method:

```
def getText(self):
    return RPython.vectorR(self.lines, "character")
```

that returns a character vector to R:

```
> pythonGet(last$getText())
[1] "Let four captains"
[2] "Bear Hamlet, like a soldier, to the stage;"
[3] "For he was likely, had he been put on,"
[4] "To have proved most royally: and, for his passage,"
[5] "The soldiers' music and the rites of war"
[6] "Speak loudly for him."
[7] "Take up the bodies: such a sight as this"
[8] "Becomes the field, but here shows much amiss."
[9] "Go, bid the soldiers shoot."
```

There are frequently a number of options in these situations. Computations on the R side could coerce the list to a specific type, if the desired type was known. The field in the Python object could itself have class "vector_R", but this would be a complication for Python computations on the field.

# Chapter 15

# An Interface to Julia

## 15.1 R and Julia

This chapter describes an interface from R to computations in the Julia language, implemented in the XRJulia package and following the XR structure described in Chapter 13. The interface is described as it would be used in an application project, via an application package that incorporates computational techniques integrated into R but using functions and/or data types implemented in Julia. The XRJulia package, the XR package it imports and the juliaExamples package are available from github.com/johnmchambers.

Julia is described as a "high-level, high-performance dynamic programming language for technical computing".[1] Its intended applications focus on computational methods for numerical, scientific and similar applications. Its design combines high-level programming structures with efficient code, compiled on-the-fly from Julia language source code.

The language and user environment are quite similar to R in many respects. Interacting with Julia, one types in expressions; the system computes results and prints output back. Function definitions and calls are very much the heart of programming with Julia. Many of the base functions and operators closely resemble those in R.

A particularly strong similarity, not shared with many other languages, is that Julia, like R, implements functional OOP: generic functions with methods selected according to the classes of one or more of the arguments in the call. Classes are called "types" in Julia and the system for type definition works differently; in particular, the use of macro-like templates for type and method definition is a key

---

[1] http://julialang.org

feature. But type definitions are objects, including the specification of properties, as in R.

The interface provides direct analogues to Julia function calls and other computations through methods for evaluator objects from the `"JuliaInterface"` class or through equivalent R function calls (Sections 15.2 and 15.3).

An application using the interface can define proxy functions in R that call corresponding functions in Julia, including the functional methods defined for these functions (Section 15.4). Proxy classes in R can be defined corresponding to types in Julia, with access to fields consistent with reference classes in R (Section 15.5). Julia does not have encapsulated methods.

Julia emphasizes a form of functional *computing*, suggestive of the ⌐FUNCTION⌐ principle, but the design is not related to functional *programming*, in the sense of protecting against side effects. Arguments are passed as references and in this sense Julia types are more analogous to reference classes than functional OOP classes in R.

In some sizable collections of functions (for example, some graphics applications) function values are largely irrelevant, with the side-effects of the function call on external objects being the main point.

Nevertheless, for "programming in the small", programming with Julia is largely based on defining functions. These are easy to define in the language and are immediately available for use:

```
function myMean(x)
   sum(x)/length(x)
end
```

This defines the function and assigns it as `"myMean"`.

Functions are generic by default. Julia has optional typing; by not declaring the type of the argument to `myMean()`, we essentially define a default method. Definitions of functions with the same function name but with explicit type declarations are the equivalent of method definitions. Argument names are arbitrary in methods; there are no formal arguments.

For medium-scale programming, Julia has packages (collections of source code) and within a package *modules*, which are declarations surrounding a collection of source code. As in other languages, modules can be imported by various mechanisms and used to define the namespace for new applications.

Applications using the interface are likely to define some Julia functions and types in module(s) associated with the package. These may usefully define methods for existing functions, simply by declaring the arguments to correspond to types that will also often be defined in the application's Julia code. The interface includes facilities for importing packages and modules from Julia (Section 15.3).

Another strong similarity between R and Julia is in their treatment of R's basic vectors, matrices and arrays. As in R, Julia has taken over the essential organization of these data structures originating in Fortran. The XRJulia interface converts data with such structure into the corresponding class in the other language. In addition, there is a general data conversion mechanism following the XR structure that supports conversion of arbitrary classes in either language (Section 15.6).

# 15.2 Julia Computations

Computations in Julia using the XRJulia interface are carried out by a Julia interface evaluator, an object from the `"JuliaInterface"` class. The current evaluator from this class is returned by the function `RJulia()`:

```
ev <- RJulia()
```

If no evaluator exists, one is started.

The XRJulia interface uses a connection, via a socket, to a process running Julia. By default, this will be a process on the machine running R. The Julia process is started when the evaluator is initialized, and given a startup script that tells it to accept and execute commands written on the socket by the R process.

At the time this book appears, the XRJulia interface is new, and all the examples shown here used the default configuration. The design, however, anticipates a more general use of sockets, as illustrated by the parallel package, for example. The evaluator object would be initialized to communicate with an existing socket connection to a Julia process using a similar startup script.

```
ev2 <- RJulia(connection = jCon)
```

In this case, `jCon` will be an open socket connection object; for example, to a Julia process initialized on a remote host for an R interface.

The choice of a connected rather than embedded interface was partly to illustrate this approach, given that the Python interface in Chapter 14 was embedded, but there are other advantages.

While embedded interfaces tend to be more efficient, at least in communicating between the languages, connected interfaces free the server computations from constraints on the design due to running the server language within the R process.

Since a connected interface is communicating with an independent process, there should be no constraints on the Julia computations because of the interface. Connected interfaces also raise the possibility of distributing computations across machines; for example, using a more powerful machine that you have to pay for when capacity is needed but a local process for less demanding computations.

General expressions and commands can be evaluated by the methods:

```
ev$Eval(expr, ...)
ev$Command(expr, ...)
```

In Julia, in contrast to R, not all statements can be evaluated as expressions; these statements usually have side-effects but no value, and will throw an error if called through `$Eval()`. For these the appropriate method is `$Command()`, which has the same arguments as `$Eval()` and evaluates the Julia string but makes no attempt to treat the result as an expression:

```
ev$Command("rtpi = sqrt(pi)")
```

Any piece of code that is complete and valid in Julia should be executable via `$Command()`.

The `$Eval()` and `$Command()` methods and all other methods in this section have functional equivalents `juliaEval()`, `juliaCommand()`, etc. These have the same arguments as the methods, plus an argument **evaluator=**, by default and usually the current Julia evaluator, which will be started if none exists.

For computations where no special evaluator is needed, the functional forms may be more natural looking in R and avoid explicit reference to the evaluator object. They do nothing but call the corresponding method.

In these methods, **expr** is a character string to be parsed and evaluated by the Julia evaluator. Additional arguments are objects that will be inserted into the expressions corresponding to C-style "%s" fields in the string. These may be results previously computed through the interface and returned as proxies for the Julia object or R objects, which will be substituted as a string that evaluates to the Julia equivalent of the R object:

```
> y <- juliaEval("reverse(%s)", 1:5)
> y
Julia proxy object
Server Class: Array{Int64,1}; size: 5
> juliaEval("pop!(%s)",y)
[1] 1
```

Scalar results are usually converted back to R values; more extensive or structured results are assigned in Julia and returned as proxy objects.

Objects can be explicitly sent to Julia and got back by the `$Send()` and `$Get()` methods and their functional equivalents. In both directions the computations rely on some conversions between objects in the two languages, as we'll consider in more detail in Section 15.6.

```
> juliaGet(y)
[1] 5 4 3 2
> x <- matrix(rnorm(1000),20,5)
> xm <- juliaSend(x)
> xm
Julia proxy object
Server Class: Array{Float64,2}; size: 100
> xjm <- juliaGet(xm)
> all.equal(x, xjm)
[1] TRUE
```

Julia has a full set of typed arrays, differing in details from R but very naturally mapped to R arrays. Essentially no information is lost in transferring numerical matrices, as in this example.

Julia operates with its own version of the FUNCTION principle; most interesting computations are done by functions. Defining new functions and/or new functional methods is the central step in programming-in-the-small, just as in R. Care is needed because this is functional *computing* rather than functional programming: functions frequently have side effects. In the examples above, for instance, the call to pop!() altered the object y.

For convenience, a function call has a short-cut for $Eval() that avoids messing with format strings. The first argument is the character string name of the Julia function, the remainder the arguments to the call. The expression to call the function pop!() above could have been written:

```
juliaCall("pop!",y)
```

juliaCall() or the $Call() method may be useful if the function name is computed rather than a constant or if the function is only called in one instance. Otherwise it's usually more convenient to define proxy functions in R, as discussed in Section 15.4.

As an alternative to a separate call to $Get(), the $Eval() and $Call() methods have an optional argument, .get=, that can be used to force conversion of an arbitrary result by supplying it as .get = TRUE.

```
> xt <- juliaCall("transpose",xm, .get=TRUE)
> dim(xt)
[1]   5 20
```

The proxy functions in Section 15.4 also have a .get argument with the same interpretation.

# 15.3   Julia Programming

As with R packages and Python modules, the names to refer to functions and other objects in Julia are organized by *modules*. In all three languages, the result is the ability to refer to objects by a fully qualified form, `package::name` in R and `module.name` in the other languages.

All three languages take slightly different routes to the organization of source when the evaluator is searching for a module or equivalent. In R and Python, the package/module is defined by the directory and file structure of the package. In R a package is a directory structured as we discussed in Section 7.1. In Python a module is a single file of source. Julia has both forms, in its own style.

Julia, like R, has the notion of a package within which directories and files are organized according to a particular structure. However, the Julia evaluator will also recognize separate files of source code through the file suffix `".jl"`. Either way, the directory or file must contain a matching `module` declaration in a particular form in the source code.

The XRJulia package, for example, has a file and module containing the various functions and data types used through the interface. This could have any name; for simplicity we use the package name and put the code in the file

```
inst/julia/XRJulia.jl
```

in the package source. This file declares the module of the same name:

```
module XRJulia
... # all the Julia source code
end
```

The evaluator will look for packages or files in one of a list of declared directory locations, analogous to `.libPaths()` in R. An application package that contains any Julia modules of its own will need to make these available by calling the function:

```
juliaAddToPath(directory, package)
```

In general, this will add any named directory to the search path, from the specified R package, or a directory unrelated to any R package if `package=""`. The `directory` is interpreted relative to the installation directory of the package. A package can refer to its own installation directory by omitting the `package` argument. If the package follows the XR convention of putting the files of Julia code into a directory `"inst/julia"` in the package source, that directory can be added to the search list by the empty call:

```
juliaAddToPath()
```

Importing modules not associated with an R package may raise difficulties for portability; see Section 13.4, page 270 for some comments.

Once a module is accessible by being on the search path, it must be imported to make its objects available by reference. As in R, there are some base objects always available, including the standard library. Objects from other modules need to be made available by importing; unlike R, the fully qualified reference will not load the module automatically.

The interface function

```
juliaImport(module, ...)
```

generates suitable `"import"` commands in Julia. Fully qualified imports are provided by calling `juliaImport()` with only the module name. To use the name in an unqualified form, supply it explicitly as a separate argument. If we wanted to use the `undigit()` function in Julia module Digits:

```
juliaImport("Digits") # Julia calls to Digits.undigit()
juliaImport("Digits", "undigit") # Julia calls to undigit()
```

Julia actually has two directives, `using` and `import`, that behave somewhat differently. The main difference is that `using` makes all the exported names available in unqualified form. The XRJulia package supports the equivalent range of options using the argument list to `juliaImport()` See the online documentation for details.

The functions and methods in the previous section were essentially equivalent, but for the path and import operations in this section the functional version is preferred. A call to `juliaAddToPath()` from the source code of a package adds the directory to a table of the path lists for all interfaces. Similarly, the `juliaImport()` function adds to a table of import directives. The functional form will add the path and module information to the current `"JuliaInterface"` evaluator and to all future evaluator objects.

For application packages, the functional form is preferred, except in the unusual case where one evaluator object needs its own path or imports. Then the method would be used for that evaluator explicitly.

Modules and source files are distinct concepts in Julia, even though a module can correspond to a single source file. The function `juliaSource()`, or the interface method `$Source()`, parses and evaluates the code in a specified file, using the Julia function `include()`.

The Julia commands `require` and `reload` also evaluate the contents of a specified file, but they put their results into the main module, meaning that assignments in the file will not be visible from interface expressions: `juliaSource()`

works through the evaluator so that all results are stored in the same module as other interface computations.

Julia, like R, returns the value of the last expression computed in the file. The following little source file defines a Julia type and returns an example object from it by calling the generator function:

```
type testT
    x::Array{Int64,1}
    y::ASCIIString
end

testT([1,-99,666],  "test1")
```

Assuming that this is file "testT.jl" in the current working directory of the R process, we can use it to compute a proxy for the object returned by testT():

```
> xt <- juliaSource("testT.jl")
> xt
Julia proxy object
Server Class: testT; size: NA
> juliaGet(xt)
R conversion of Julia object of composite type "testT"

Julia fields:
$x
[1]    1 -99 666

$y
[1] "test1"
```

The field names for an object of type "testT" are used in evaluating the juliaGet() call to create an Robject of class "from_Julia". Data conversion techniques will be discussed in Section 15.6.

In developing software for the interface, it may be convenient to have Julia print some result, rather than having to convert that to an R object and bring it back. For this purpose, you can use the juliaPrint() function. It can take one argument, typically a proxy object. Or, you can give it several arguments that will be interpreted as if given to the $Eval() method and will print the result of that computation. See page 337 for an example.

# 15.4  Julia Functions

A proxy function in R to call a Julia function of a specific name is returned by

```
JuliaFunction(name, module)
```

with the argument `module` only required to ensure that the corresponding module is imported, if it is not by default. As an example, the Julia function `svdfact()` computes a singular value factorization of a matrix. An R proxy function for it could be created by:

```
svdJ <- JuliaFunction("svdfact")
```

Calls to `svdJ()` will generate calls through the XRJulia interface to `svdfact()`. The arguments to `svdJ()` will be converted as needed from the R objects; more likely, except for simple scalars, they are R proxies for Julia objects previously computed or converted. By default, the evaluator used is the current Julia evaluator, which will be started if necessary; an optional argument allows a different evaluator to be specified. The `svdfact()` function is part of the standard library, so no explicit module import is needed.

We can construct the decomposition of the Julia array `xm` shown on page 325:

```
> sxm <- svdJ(xm)
> sxm
Julia proxy object
Server Class: SVD{Float64,Float64}; size: NA
```

The composite Julia type for the result has essentially the same information as the result of the R function `svd()`. In Section 15.5, proxy R classes for the type will be shown. With or without a proxy class, the interface evaluator can get the information in the decomposition back to R, as we will show in Section 15.6.

Julia functions are generic by default; that is, a function definition actually creates a method associated with that function name. Optional type declarations for the arguments specify the signature for the method. Additional function definitions for the same name, with different type declarations for the arguments, will define additional methods for the generic.

As with R, the classes (types) of the actual arguments in a call will be used to select the best method for that call. An important difference, however, is that Julia uses the selection to compile the appropriate method for this case.

This distinction has some implications for an interface from R. There is no formal argument list for the function, and indeed no pre-determined number of arguments. Different methods may have different argument names or number of arguments.

Julia methods are best seen as prescriptions for creating (by compilation) an actual executable method. Method "dispatch" examines the type declarations for existing methods to find a match to the types of the actual arguments. The `svdfact()` function, for example, is implemented for (currently) 9 signatures corresponding to the function declarations:

```
svdfact(D::Diagonal, thin=true)
svdfact(M::Bidiagonal, thin::Bool=true)
svdfact{T<:BlasFloat}(A::StridedMatrix{T};thin=true)
svdfact{T}(A::StridedVecOrMat{T};thin=true)
svdfact(x::Number; thin::Bool=true)
svdfact(x::Integer; thin::Bool=true)
svdfact{T<:BlasFloat}(A::StridedMatrix{T}, B::StridedMatrix{T})
svdfact{TA,TB}(A::StridedMatrix{TA}, B::StridedMatrix{TB})
svdfact(A::Triangular)
```

Many of these are templates; that is, specific argument types will match the signature for some macro-style substitution of the template argument, such as T, TA, TB in the methods above. It is part of the central Julia design that this provides a flexible, dynamic method selection system. It would not be straightforward, however, to check on the R side that the arguments to the proxy are consistent with the available methods.

Since argument names are not restricted by the generic function, as the example shows, function calls in Julia cannot refer to arguments by name. Julia does provide a mechanism for "keyword" arguments. These are defined by a special syntax in the formal argument list for a particular method; in effect, they match elements in a dictionary to keyword arguments in the call. But ordinary arguments are accepted only positionally.

Considering these characteristics, the present version of the XRJulia interface leaves argument checking up to the server language side of the interface. Proxy functions in R for Julia functions pass the actual arguments on unmodified. The number and order of actual arguments should be what is intended for the Julia call and named arguments will be passed on with the same names. Note that named (aka keyword) arguments must follow positional arguments in the call.

If the `module` argument is specified in the call to `JuliaFunction()`, the named function is assumed to be exported from that Julia module. The body of the proxy function will include an import call for the module; because the XRJulia evaluator keeps a table of imported modules, only one actual import command will be issued to Julia. The actual Julia function call uses a fully qualified name; therefore, proxy functions can interface to two functions of the same name in distinct modules.

# 15.5 Julia Types

Julia provides for definitions of what are called "composite types" and are in effect classes with specified fields. Since Julia supports a form of functional OOP, these are used more as functional classes in R. They appear in Julia as type declarations in method definitions, analogous to the signatures for R methods. They do not have encapsulated methods, in contrast to classes in Python or Java.

Unlike functional class objects in R, Julia objects use reference semantics; when you change a field in a Julia object the change is not local to the function call where it takes place.

A call to `setJuliaClass()` creates a proxy class in R for a type in Julia:

```
setJuliaClass(juliaType, module)
```

The arguments are the type name in Julia and the module name, which can be omitted for classes in the base software. Metadata in Julia defines the fields, which will be accessible as reference class fields in R, using the `` `$` `` operator.

Many relevant Julia types are *parametrized*, in that their definition contains one or more template- or macro-style arguments. In the example on page 329, the result returned was a proxy for an object of type `"SVD{Float64,Float64}"`. The `"SVD"` type is parametrized by (at least) two numeric types, for the input data and the output values.

When the type is parametrized, either the specific version or the whole family may have a proxy class defined in R. The field names of the class are generally defined by the family, with only the field types affected by the specific type; however, it may be undesirable in R to use the same proxy class for all the specific types. When a proxy object is returned from Julia, the XRJulia interface looks first for a proxy class to the parametrized type and then for the unparametrized version. The application package can choose which version to set up, or both. In the example:

```
setJuliaClass("SVD")
```

would create a proxy class for any member of the family; all have the same fields, `"U"`, `"S"` and `"Vt"`. If this proxy class had been defined before setting up the proxy function `svdJ()`:

```
> sxm <- svdJ(xm)
> sxm
R Object of class "SVD_Julia", for Julia proxy object
Server Class: SVD{Float64,Float64}; size: NA
> sxm$S
```

```
Julia proxy object
Server Class: Array{Float64,1}; size: 5
```

Julia types have the additional option of being "immutable"; effectively, this means that all the fields are read-only in the sense discussed in Chapter 11. Their fields may be accessed but not assigned. Having the relevant fields read-only may be a useful way to avoid accidental invalidation of the object when fields must have a fixed relationship. Such is definitely the case with "SVD" and other matrix factorizations; manipulating values in any of the fields will usually invalidate the object as a correct factorization. And in fact the "SVD" type is declared immutable.

If XRJulia detects an immutable type, it makes the proxy fields read-only.

```
> sxm$S <- 0
Error: Server field "S" of server class "SVD{Float64,Float64}"
             is read-only
```

## 15.6   Data Conversion

Data conversion in XRJulia is based on that described in Section 13.8 for XR, including facilities for representing general objects from R and from Julia, but provides additional features that may be particularly relevant for numeric and other algorithmic interface applications.

R and Julia have a number of similarities in the representation of important classes of data, particularly those corresponding to vectors, matrices and arrays in R. There is also a natural relation between classes in R and composite types in Julia. Data conversion in XRJulia uses these characteristics for a cleaner and more direct matching between the languages than provided by the default strategy. You can usually assume that objects map automatically in both directions if they come from classes that are vectors or arrays of any of the types known in R or that consist of slots/fields that can themselves be mapped.

Applications can customize the conversion when that is helpful. Conversion to Julia implements methods directly for asServerObject(), omitting the JSON intermediate form. Conversion to R (page 335) uses methods for the generic functions toR() in Julia and asRObject() in R. We'll note limitations for converting some Julia types. Applications can often work directly from the general representation forms for the corresponding classes in the other language. An example using the R "data.frame" class is on page 337; a small example with a Julia class is in Section 15.3, page 328.

## Vectors and Arrays in R and Julia

The two languages share an approach to arrays. In both languages, arrays are defined by a block of elements of a particular type; in other words, a `"vector"` in R terminology. This vector is interpreted as a $k$-way array by associating with it $k$ integers for the range of indices in each dimension. In R the concept is implemented as the `"array"` class with slots for data and dimensions. Julia has a parametrized set of types, without explicit fields for data and dimensions but with a paradigm for programming that supports essentially the same range of objects.

In addition to the matching of array structure between the languages, R and Julia support a variety of basic data types for arrays, as opposed to the JSON notation, which only supports lists of arbitrary elements. Conversion between corresponding classes is automatic in both directions provided the basic Julia type corresponds to one of the R vector types.

Julia has a set of parametrized `Array` types

```
Array{T,N}
```

where `T` is a Julia type corresponding to the type of the elements and `N` is the number of dimensions.

Vectors in R map into one of the `Array{T,1}` types, with `T` determined by the R type of the vector. The type parameter `T` has a variety of options, considerably more than the range of basic types in R. Integer, floating point and bit-string types have options for length; R maps `"integer"`, `"numeric"` and `"raw"` into particular choices that reflect the R implementation. Julia type `"Any"` corresponds to type `"list"`. Sending vectors of various types from R will create suitable Julia array objects:[2]

```
> ev$Send(1:3)
Julia proxy object
Server Class: Array{Int64,1}; size: 3
> ev$Send(c(1,2,3))
Julia proxy object
Server Class: Array{Float64,1}; size: 3
> ev$Send(c("red","white","blue"))
Julia proxy object
Server Class: Array{ASCIIString,1}; size: 3
> ev$Send(list("Today", 1:2, FALSE))
Julia proxy object
Server Class: Array{Any,1}; size: 3
```

---

[2]In this section, we are looking at implementation details and will revert to showing the method version of `$Send()`, etc., rather than the equivalent functions.

The Julia server language expression is the list of elements, written out explicitly:

```
> ev$AsServerObject(1:3)
[1] "[1,2,3]"
> ev$AsServerObject(c(1,2,3))
[1] "[1.0,2.0,3.0]"
> ev$AsServerObject(c("red", "white", "blue"))
[1] "[\"red\",\"white\",\"blue\"]"
> ev$AsServerObject(list("Today", 1:2, FALSE))
[1] "{ \"Today\", [1,2], false }"
```

Julia interprets the list as an array of the type needed, similar to the `c()` function in R, except that elements of length > 1 effectively force a list-style object.

Complex is not a basic type in Julia but essentially a parametrized type for representing pairs of values. The `"complex"` vector in R corresponds to one of those, for pairs of floating point numbers.

```
> cx
[1] 7.8+3.8i 5.5+3.4i 5.2+3.1i 6.8+0.1i
> cxj <- ev$Send(cx)
> cxj
Julia proxy object
Server Class: Array{Complex{Float64},1}; size: 4
> ev$Get(cxj)
[1] 7.8+3.8i 5.5+3.4i 5.2+3.1i 6.8+0.1i
```

Complex vectors in R are sent by a call to the generator function for the Julia type:

```
> ev$AsServerObject(cx)
[1] "complex([7.8,5.5,5.2,6.8], [3.8,3.4,3.1,0.1])"
```

The Complex types in Julia have a generator with two vectors for the real and imaginary parts as arguments.

An R array object will also map to one of the Julia parametrized array types. For example, the `iris3` object in the datasets package is a three-way array:

```
> dim(iris3)
[1] 50  4  3
> typeof(iris3)
[1] "double"
```

Sending this object to Julia produces a corresponding Julia array object:

```
> irisJ <- ev$Send(iris3)
> irisJ
Julia proxy object
Server Class: Array{Float64,3}; size: 600
```

A general array object in R is sent to Julia by first creating the one-way array with the data part and then using the Julia function `reshape()` to specify the dimensions:

```
> xm <- matrix(1:6,3,2)
> ev$AsServerObject(xm)
[1] "reshape([1,2,3,4,5,6], 3,2)"
> ev$Send(xm)
Julia proxy object
Server Class: Array{Int64,2}; size: 6
```

## Converting **Julia** objects

The conversion of Julia objects to R retains JSON notation in the string returned by the Julia evaluator to R. Where the Julia type has a matching R class, the JSON form uses the representation of a general R object by a specialized dictionary containing an element named `".RClass"`. The conversion produces an object of the corresponding R class. R methods for `asRObject()` may further specialize conversion of this object.

Two special R classes are particularly important: `"vector_R"` and `"from_Julia"`. The first of these explicitly represents various types of vectors in R, which would otherwise be ambiguous if written as just a JSON list. The second explicitly identifies a Julia object from a composite type, converted with a named list of the (converted) data in each of its slots. This representation is not dependent on the existence of a proxy class in R for the Julia type.

The Julia side of the interface consists of a collection of methods for the function `toR()`. Its argument is an arbitrary Julia object and it returns another object such that the JSON representation produces an R object matching the original Julia object.

Objects from the parametrized `"Array{T,N}"` types are returned as R vectors or arrays. The returned object will be a vector if N is 1 and an array otherwise. The type of the R vector will be numeric, integer, logical or character for T a corresponding Julia scalar type. Type `Any` will be returned as a list. Returned arrays are constructed by reshaping the array into a one-way array and converting this for the `".Data"` slot; therefore, the same type matching applies as for vectors.

Dictionaries will be returned as named lists of their elements. While Julia dictionaries are parametrized by the type for the keys and the type for the elements, named lists imply character string keys. JSON dictionaries also require strings as keys, so the necessary coercion has already taken place to produce the JSON string.

Scalars of the types recognized by JSON will turn into vectors of length 1 of the corresponding R vector class.

Putting all this together, the convertible Julia objects include all:

1. Scalars, arrays and dictionaries; and

2. Composite types

provided that the elements of the arrays and dictionaries and the fields of the composite types are themselves convertible objects. Any such object will be converted by the `$Get()` method or by a proxy function with `.get = TRUE`, to an R object of the simple forms described above.

Application packages may want to specialize the object returned, to generate a particular R class or perform some transformation of the fields in the Julia object. The most natural approach is to write one or more functional methods in Julia for the function `toR()`. In the Julia code for XRJulia, `toR()` takes as its argument the Julia object that results from evaluating an expression or command. The value returned by `toR()` should be the Julia object that will be converted to JSON representation and sent to R.

Two-way array objects, for example, will be turned into a dictionary with an explicit R representation for class `"matrix"`. The Julia function `RObject()` produces the explicit representation; the method for array objects will finish by calling `toR()` again with the value returned by `RObject()`.

Application methods are likely to do something similar, transforming the Julia type into a chosen class of R objects. Take a look at methods for `toR()` in the XRJulia package, in file `"julia/XRJulia.jl"`.

Occasionally, the application may need to do some further computations on the particular class of R objects returned in this way, by defining a method in R for the function `asRObject()`. In both R and Julia you need to import the generic function `asRObject()` or `toR()` into the application package in order to define methods for it.

Some current limitations on conversions are due to basic types that do not correspond between the languages. Julia has a range of parametrized scalar types that have no direct R equivalent; it's unclear how important these may be for data-based applications, but some extensions to the XRJulia facilities may address these types in the future.

Other types in Julia are useful as programming steps but have a transient form that doesn't survive current conversion computations. In the example on the next page, we used `juliaPrint()` to print an object returned by the `keys()` function, rather than converting the result to R. The object returned by `keys()` is an iterator over a dictionary, a useful type in Julia. But as a composite type for conversion, it contains the entire object over which the iteration takes place, not the keys as strings.

### General R class representation: an example

The representation of a general R object as a dictionary with special keys allows computations for a class that does *not* have an obvious Julia counterpart. For an example, let's look once more at `"data.frame"`. As we discussed in Section 10.5, whether formally defined or not, `"data.frame"` effectively extends `"list"`, with slots `"names"` and `"row.names"`, equivalent to:

```
setClass("data.frame",
    slots = c(names = "character",
            row.names = "data.frameRowLabels"),
    contains = "list")
```

Julia has no type directly corresponding to this: It's essentially a dictionary, constrained by requiring the elements to represent variables with the same number of observations, plus a field for the row names. We could define such a composite type, but currently there is not much that can be done with it. More likely, a data frame sent from R will be the source for derived matrix objects, as it often is in R.

The conversion to Julia therefore uses the dictionary representation for a general R class. Section 13.8, page 289, showed an example in JSON notation, for class `"ts"`. The Julia dictionary form is similar. Let's look at a small sample from the data frame version of the `"iris"` data:

```
> iSample <- iris[sample(150,6),]
> jSample <- juliaSend(iSample)
> jSample
Julia proxy object
Server Class: Dict{Any,Any}; size: 7
> juliaPrint("keys(%s)", jSample)
{".type","names",".Data",".RClass",".extends","row.names",
    ".package"}
```

Elements `".type"`, `".extends"` and `".package"` further describe the object's class. All other elements are the slots of the R object, converted to Julia. The `".Data"` element is a list (type `"Array{Any,1}"`) of the 5 variables in the data frame.

Assuming some Julia computations modified this object or created a similar one, getting it back will create the correct R object. As we can test:

```
> iSampleBack <- juliaGet(jSample)
> all.equal(iSampleBack, iSample)
[1] TRUE
```

It's important that this works because of methods in both R and Julia, but *not* methods for the specific "data.frame" class in either case.

In the to-Julia direction, the relevant method is for asServerObject():

```
> selectMethod("asServerObject",
+               c("data.frame", "JuliaObject"))
Method Definition:

function (object, prototype)
{
    attrs <- attributes(object)
    if (is.null(attrs) || identical(names(attrs), "names"))
        .asServerList(object, prototype)
    else .asServerList(XR::objectDictionary(object), prototype)
}
<environment: namespace:XRJulia>


Signatures:
        object        prototype
target  "data.frame"  "JuliaObject"
defined "list"        "JuliaObject"
```

The method is inherited from the "list" method. If the object was simply a list, with or without names, it would be sent directly as a dictionary or array in Julia. But the method checks for additional attributes which will always be there for a class that extends "list", such as "data.frame". If so, the object converted, from objectDictionary(), will turn into a Julia dictionary with an element having the reserved name ".RClass", in this case containing "data.frame".

The converted slots will be in elements of the dictionary with the slot names. Julia computations designed for the imported R object could modify these elements or construct new Julia objects with the same structure.

Coming back to R, the object will start as a dictionary in JSON. This turns into a list, with names. The asRObject() method for "list" checks for ".RClass" among the names; if found, an object from that class will be constructed.

# Chapter 16

# Subroutine Interfaces;
# An Interface to C++

## 16.1   R, Subroutines and C++

The XR interface structure described in Chapter 13 models the interface in terms of a server language evaluator. The user of Python or Julia, for example, interacts with an evaluator that parses and evaluates expressions. Subroutine interfaces are fundamentally different and the XR structure does not apply.

The goals of the XR structure still make sense, however, particularly *convenience* and *generality*. An application package extending R would like its users to be essentially unaware of the interface; functions and objects should seem natural in R.

The programming for the application should be convenient and general as well. Two of the main tools for the XR structure would be equally relevant for subroutine interfaces: functions and classes in R that are proxies for those in the server language.

As with the XR interfaces, we would like general server-language computations to be accessible and to allow general R data to be used as arguments or produced as results.

Fortunately, there is an existing and widely used package, the Rcpp interface to C++, that goes a substantial way towards achieving these goals, including proxy functions and classes and a general approach to data conversion. Using Rcpp to extend R is the main topic of this chapter. Rcpp is available from CRAN.

# 16.2   C₊₊ Interface Programming

The interface to C++ provided by the Rcpp package can be helpful in extending R in several scenarios, requiring a varying amount of actual C++ programming.

1. If the goal is to create an interface to an existing C or C++ function not originally written for use with R, Rcpp can automate much of the interface programming. If the data types for the arguments and results are among those already accommodated by Rcpp, both a C++ and R function for the interface can be created with one setup step and no C++ programming.

2. There exists a large body of software in the form of C++ libraries, including some high-quality, potentially useful computations. Interfacing to such libraries usually requires identifying or developing C++ functions and possibly C++ classes to use from R.

3. C++ may be the language of choice for a new programming project, particularly if an interface with R is one of the goals. C++ is not as friendly for relatively casual computing as, say, Python or Julia, at least in my opinion. But modern C++ is a powerful language capable of producing efficient code for a range of applications.

The scenarios are not mutually exclusive. A substantial project ("programming in the large" in my terminology) may at various stages involve all three.

### Examples

We will look at an example of each scenario in this chapter. For an interface to a C++ function, we will look again at the standard, if not very realistic, `convolve()` example considered in Section 5.5, as a C++ function (page 345).

For an example that emphasizes an interface to an existing library, we consider creating an interface from R to the JavaScript language following the XR structure of Chapter 13. This project would create a package similar to XRPython and XRJulia, likely to be named XRJavaScript unless that was considered too long.

The relevant C++ library is V8 (or Chrome V8), an open-source C++ library and associated software provided by Google as a "JavaScript engine". It has been used in the CHROME web browser (`https://developers.google.com/v8`) and a number of other applications. V8 has an interesting design, taking advantage of C++ features. The role of V8 in the web browser and in other projects makes one reasonably confident that it will continue to be supported, evolving and well tested.

A subroutine interface to the V8 library is an attractive tool to build an XR interface to JavaScript. Better yet, it's largely been done already. The package of the same name, V8 [28], implements an interface from R to JavaScript. This is not an XR interface but has a number of similar features and includes, as we will see, some key subroutine interfaces to support such a project. (The V8 and XR packages were implemented independently.)

As an example of using a subroutine interface, implementing a language interface may seem a bit bizarre, but in fact it follows the [INTERFACE] paradigm quite closely. We have a substantial project in mind and some essential requirements: evaluating JavaScript expressions, sending the input data and getting the results back to R. Existing software (the V8 library) carries out much of what we need, but in another language (C++). The principle encourages us to examine an interface to that software as a contribution to our project.

We'll need to distinguish the Google V8 project and its associated C++ library (our main focus) from the R package of the same name; to do this, I will use the lower-case "v8" to refer to the library (as it would appear in a directory name). To be continued (page 346).

For an example of the third scenario, we will look again at implementing an encapsulated OOP class in R that models an evolving population. In Section 9.5, page 143, we used a very trivial model to illustrate such classes in R. The population consisted of identical individuals, each having a given probability of dying or of splitting into two at each generation.

Such a model suits C++ also, perhaps even more naturally, and if one were looking to run such simulations on a large scale, the greater efficiency of C++ might be relevant. Rcpp provides some handy tools for the purpose, one of the most important being its integration of random number generation in C++ into the R paradigm, providing a more consistent and comprehensible statistical result than most ad hoc simulation in C++.

We will also add one feature to the earlier model. Instead of identical individuals, the population starts off with $n$ individuals, each having its own birth and death probability. So the population can be defined by two vectors, `birth` and `death`.

The population evolves as before, but each individual dies or gives birth according to its own probabilities. A new individual inherits these probabilities from its parent. We define a single evolutionary step as first a death process, randomly killing off members of the population according to their death probabilities. Then the survivors get a chance to divide, adding a cloned member to the population (with the same birth and death probability as its parent).

The computations as implemented are highly iterative and depend on being able to grow or shrink the vector representing the population in response to births or deaths. Simple iterative computations are of course much faster in C++ than in R. Also, vectors that can grow and shrink, without being copied each time, are exactly what the C++ standard template library is designed to implement. If we were to worry about efficiency, running the evolution of a population is likely to be much faster in C++ using the methods for vectors in the standard template library.

Here is one potential definition for such a C++ class:

```cpp
#include <Rcpp.h>
using namespace Rcpp ;

class PopBD {
  public:
    PopBD(void);
    PopBD(NumericVector initBirth, NumericVector initDeath);

    std::vector<double> birth;
    std::vector<double> death;
    std::vector<long> size;
    std::vector<int> lineage;
    void evolve(int);

};
```

The class has four fields—the two vectors of `birth` and `death` probabilities, a vector of the population size after each evolutionary event, and a vector, `lineage`, that identifies the original population member that was the ancestor of each current member. There are two constructors, the relevant one taking the birth and death probabilities as numeric vectors from R. The only method is `evolve()`, which evolves through a specified number of generations of the death-birth procedure. Individuals that die are removed from `lineage` and then any new births are added by appending the corresponding index for the original ancestor to `lineage`. We'll consider this class in the context of interfacing a C++ class to R, on page 349. For an actual implementation, see the XRcppExamples package at `github.com/johnmchambers`.

# 16.3  C++ Functions

Proxy functions are a powerful mechanism for interface programming: a function is called in R that is in effect a call to a corresponding function in the server language. Rcpp provides a straightforward mechanism for creating proxy functions through C++.

We begin with a target C++ function, which we would like to call from R. As usual in C++, by "function", we mean a group of statements given a name, defined with a sequence of arguments. Each of the arguments and the named function itself has a declared type. The function has to be accessible from R, either as part of the source of the application package or from some available C++ library.

Although we're calling this a C++ function, only the outer layer needs to be C++, which has a mechanism for declaring an external name to come from C. The proxy function interface described here is therefore an alternative to the basic .Call() interface. But in contrast to that interface, the target function will *not* be required to have pointers to R objects as the types for its arguments or return value.

For each target C++ function two new functions are generated, a C++ interface function and an R interface function.

- The C++ interface function is suitable to be called via the .Call() interface from R. Its arguments correspond to those of the target C++ function, but the arguments and the return value of the interface function are all declared **SEXP**, as .Call() requires. The interface function converts each argument into an object of the declared type of the target argument, calls the target function and returns its value, converted to type **SEXP**.

- The R interface function takes the same set of arguments and returns the value of a .Call() to the C++ interface function. By default, it has the same name and the same argument names as the target C++ function.

Users of the package then have an R function with the same calling sequence as the C++ function; a call to the R function returns the value of the corresponding C++ function call as an R object.

The C++ and R source code for the interface functions are created before the package is installed, by what we've called a *setup step* (Section 7.1, page 112). From an R process, using the Rcpp package:

```
> compileAttributes(pkgdir)
```

where **pkgdir** specifies the package source directory. As its essential side-effect, **compileAttributes()** creates a file of C++ code and a file of R code, each of

which is written into the appropriate subdirectory of `pkgdir`.[1] All the R proxy
functions for the package are written to a single file, and similarly for the C++
functions.

For the Rcpp interface to work, the essential requirement is on the data types
of the arguments and of the returned value for the C++ function. There must be
a definition for the conversion of the R objects supplied as actual arguments to
the declared C++ data types and similarly for the conversion of the data type for
the function's value to an R object.

Rcpp has a large and growing range of classes that are recognized. Most importantly, the conversion and as a result the whole attribute interface is built
on a template programming mechanism in C++ that can be extended for new
applications.

Two examples will illustrate, first one with a C++ function written for use
with R, then with some functions calling an existing C++ library.

## Example: C++ for R

Starting in Section 5.5, page 90, we used the by-now-standard `convolve()` function
in discussing efficiency and vectorizing. We showed there a very simple C function
and a corresponding R function, totally unsuited to R and much slower. The C
version can be written in C++. Here's a version taken directly from the Rcpp
vignette describing attributes [1].

```
#include <Rcpp.h>

using namespace Rcpp;

// [[Rcpp::export]]

NumericVector convolveCpp(NumericVector a, NumericVector b) {

    int na = a.size(), nb = b.size();
    int nab = na + nb - 1;
    NumericVector xab(nab);

    for (int i = 0; i < na; i++)
    for (int j = 0; j < nb; j++)
            xab[i + j] += a[i] * b[j];
```

---

[1]The name `compileAttributes()` refers to C++ attributes, but these are not actually used and anyway have nothing to do with R attributes.

```
return xab;

}
```

The C++ function is a simpler version of the C version shown on page 94, using C++ classes to avoid clumsy macros and not requiring `SEXP` types.

The odd-looking comment line containing `"Rcpp::export"` flags the following function as a target for an interface from R. For simple examples, just treat it as an incantation that will always precede a function or function declaration when we want an R proxy function to be generated.

In the example, the arguments and return value are all declared `NumericVector`, a C++ class specially defined in `Rcpp` to correspond to `"numeric"` vectors in R. Conversions from and to R essentially use the data part of an R object as an array in C++. The class has some methods, such as `size()`, that correspond to R vector properties. Generator functions create new objects, in this case `xab` which will be returned as the value. The class also has C++-style methods and overloaded operators that let it behave like a similar C++ array, as in the only actual computation in the function:

$$xab[i + j] \mathrel{+}= a[i] * b[j];$$

For efficiency, the key point is that these are C++ operations without the overhead of R functions.

As usual, our assumption is that this function comes as part of an R package, in this case named XRcppExamples. The setup step for this package will call the `compileAttributes()` function, applied to the source directory of the package. When that function finds the `"Rcpp::export"` comment, it proceeds to parse the following function declaration, obtaining the type and name of the function and the types and names for the arguments.

With this information, it constructs the two interface functions, in C++ and R. The type declaration of the generated C++ function is:

```
SEXP XRcppExamples_convolveCpp(SEXP aSEXP, SEXP bSEXP);
```

The names of the package and of the target function are concatenated for the interface function's name.

The R interface function is a proxy for the target C++ function normally with the same function name and argument list:

```
convolveCpp <- function(a, b) {
    .Call("XRcppExamples_convolveCpp", PACKAGE = "XRcppExamples",
        a, b)
}
```

The basic interface `.Call()` is to the generated C++ function, which then calls the target function. The files `"R/RcppExports.R"` and `"src/RcppExports.cpp"` contain all the R and C++ code generated.

The C++ interface function constructs `"NumericVector"` objects from the `"SEXP"` arguments. It then calls the target function, converts the result from `"NumericVector"` to `"SEXP"` and returns the result. There is one other step, to co-ordinate random number generation, not relevant in this case but essential in our next example.

The details of conversion will be examined in Section 16.5.

One important cautionary note: The conversion to `"NumericVector"` does not copy the R object (the same is true of the underlying `.Call()` interface). Therefore, the target C++ code should not modify any values in the arguments without doing an explicit copy. Such changes could corrupt R objects in the calling functions. Assignments should be to local objects generated in the C++ code, such as vector `xab` in `convolveCpp()`.

### Example: Interfacing to a C++ library

The `convolve()` example illustrates the Rcpp mechanism, but has nothing to do with real interface applications. For an example that does, let's return to the v8 C++ library, and look at some interfaces to that from Jeroen Ooms's V8 package. The C++ library provides useful subroutine interfaces for communicating with JavaScript.

Our original goal is to construct an XR-style interface class for JavaScript. The two key methods to specialize the R reference class are:

```
$ServerExpression(expr, ...)
$ServerEval(string, key, get)
```

(Section 13.5, page 272).

`$ServerExpression()` takes an expression involving some number of R objects (the `"..."` arguments) and substitues server-language expressions for each of these in `expr`. Chapter 13 includes a procedure for this using JSON. Given that JSON is an acronym for *Java*S*cript* O*bject* N*otation*, we can expect to take over the XR method unchanged. Let's turn to the second method.

`$ServerEval()` should parse and evaluate a server-language expression or command supplied as `string`. The XR strategy is to return the computed value converted to an R object if it is a scalar and a proxy for the result otherwise, with special values for `key` and `get` forcing conversion or causing the value to be ignored, and with condition objects being returned in all cases if an exception occurred.

The XRPython and XRJulia packages have server-language functions to implement this strategy, with the same arguments as the R method. A JavaScript interface would likely follow the same design. The R side then needs to construct and evaluate a call to that server language function.

How can the v8 library help? A process using the library starts a virtual machine ("engine") to evaluate JavaScript expressions but actual evaluation takes place within one of potentially many context's. A context object is effectively a namespace within which expressions can be evaluated and assignments recorded and accessed.

Contexts are clearly analogous to an evaluator in XR; the $initialize() method of the evaluator would create a context through a call to create the C++ object and keep (a pointer to) the result as a field in the evaluator, say "context".

This leads to our first target C++ function, to create a context. The V8 package has that:

```
// [[Rcpp::export]]
ctxptr make_context(bool set_console){
    ....
}
```

The return type "ctxptr" is a typedef that uses a powerful construction in Rcpp, the parametrized type "XPtr". This arranges for a pointer to be converted to and from R using R's "externalptr" type but incorporating information about the actual data type of the C++ reference.

With a proxy function for make_context(), we can create a context object. The next step is to use the context object to send a string expression to JavaScript for evaluation. The V8 package also has a C++ function, context_eval(), for that:

```
// [[Rcpp::export]]
std::string context_eval(std::string src,
        Rcpp::XPtr< v8::Persistent<v8::Context> > ctx) {

    .....

}
```

The arguments are src, the string containing the expression to evaluate, and ctx, the context object in which the evaluation will take place. The function returns a string representing the value computed. The "string" type is defined in the standard C++ library; Rcpp provides conversions for these types. The explicitly parametrized "XPtr" type for the context is a version of the "ctxptr" type returned by `make_context()`. We'll say more about type declarations in Section 16.5 and about `"XPtr"` on page 354.

With R proxy functions for these two C++ functions, our hypothetical XR-JavaScript project would be well on the way. How typical is this example?

Clearly, we got a "free ride" from the existing V8 package that implemented two key C++ target functions. Starting from just the v8 C++ library would have involved more learning and more programming. You can look at the source for the V8 package [28] to assess that.

For other projects, your luck will be variable. There are many hundreds of existing packages using Rcpp, however, so finding one that has relevance for some of what you need is not an unreasonable hope.

In any case, the C++ programming task can be formulated clearly: to define some target functions that carry out the relevant computations and return references to objects containing the information the application project needs. Provided the data types for arguments and results fall within the range covered by Rcpp, R proxy functions can be created automatically.

## 16.4   C++ Classes

C++ started out as "C with classes" and its encapsulated OOP nature is central to many applications using the language. It is possible to define R proxy classes to access the fields and methods of a C++ class. The function `exposeClass()` in package Rcpp constructs R and C++ source code to define a proxy for a specified C++ class. As with `compileAttributes()` for proxy functions, `exposeClass()` will be called in a setup step, writing files into the source directory for the application package. Unlike `compileAttributes()`, the details about the class are provided in the arguments to `exposeClass()` rather than being inferred from the C++ source.

The information needed for a proxy class definition is more difficult to determine than for a proxy function. The source structure is more general and class definitions often inherit from other classes, potentially imported from other libraries. The current proxy class mechanism requires the calls to `exposeClass()` to define the fields and methods for which R proxies should be created. In addition, inherited fields and methods require extra information in the call to `exposeClass()`.

Let's begin by assuming the C++ class definition does not include inherited classes. The main arguments to `exposeClass()` are `class`, the name of the C++ class to be exposed, and:

`fields=`, `methods=` the names of the C++ fields and methods to be made available;

constructors= a list, with each element the data types for the arguments to a
constructor function to be made available;

header= one or more character strings for, typically, #include directives for files
containing the class definition;

readonly= the names of any of the fields that should not be writable from R.

The call to exposeClass() should be made with the working directory set to the
top-level source directory of the application package. C++ and R files are written
to the "src" and "R" subdirectories.

As an example, let's look at the class "PopBD" that we defined on page 342
to represent a model of an evolving population. The class has four fields, two
constructors and a single method, evolve(). If the class is defined in the package
XRcppExamples, files to produce a proxy version for this C++ class would be
produced by:

```
> require(Rcpp)
Loading required package: Rcpp
> setwd("XRcppExamples")
> exposeClass("PopBD",
+        constructors =
+          list("", c("NumericVector", "NumericVector")),
+        fields = c("lineage", "size"),
+        methods = "evolve",
+        header = '#include "PopBD.h"',
+        readOnly = c("lineage", "size"))
Wrote C++ file "src/PopBDModule.cpp"
Wrote R file "R/PopBDClass.R"
```

Compare the arguments to exposeClass() with the C++ class definition. The
fields and methods arguments name those to be made available in the R version
of the class. The constructors argument is a list of the signatures for which
constructors are to be made available. The first has no arguments, the second has
two arguments both of data type NumericVector.

The C++ code generated needs to find the definition of the class; this is the
role of the header= argument, which should be a character vector of additional
lines needed in the resulting C++ file. Typically, as here, the code includes one or
more header files defining the class. The computations in exposeClass() do not
use these files, but C++ compilation will need them when the package is installed.

Exposing the one method lets us drive the evolution from R, and the two
fields exposed summarize the current state. Since the birth and death fields

came from R and are constant, the interface chooses not to export them. The two exported fields are made read-only to prevent accidentally breaking the model by assignments from R.

When the package—XRcppExample in this case—is installed, the C++ code will be compiled and linked. When the package is loaded, a load action set up by the generated R file will create a reference class and its generator—"PopBD" and PopBD()—with the appropriate pointers to the C++ class.

The R class has proxies for the constructors, methods and fields of the C++ class. As an example, an object is generated from the class with 20 individuals, 10 having each of two different birth probabilities.

```
> birth <- rep(c(.04,.06),10)
> death <- rep(.02, 20)
> p1 <- PopBD(birth, death)
> p1$size
[1] 20
> p1$evolve(100)
> table(birth[p1$lineage+1])

0.04 0.06
  72  351
```

The initial model population was created using the two-argument constructor. That and the $evolve() method do all their calculations in C++. Notice that the call to $evolve() supplied the usual R numerical value, which was coerced suitably; although the C++ argument is of data type int, one does not need to supply 100L as the argument. After 100 evolutionary steps, we can see that the population has grown by a factor of 20 and the slightly more "fertile" individuals are starting to dominate.

The C++ file written by exposeClass() defines an Rcpp module. Modules expose specific constructors, fields and methods for a C++ class in a form that can be called from R. See the rcpp-modules vignette with the Rcpp package for details. The R file written just sets up a load action that in turn calls the function loadRcppClass(). The result is to create a reference class corresponding to the C++ class in the namespace of the package at the time the package is loaded. The package can export this class or not, depending on the export directives in the "NAMESPACE" file of the package.

The exposure of a class using the Rcpp Module mechanism does not handle inherited methods or fields directly, but exposeClass() will create an interface to them, given extra information about the data types involved. Explicitly defined members can be identified simply by the name of the field or method, but the specification of members inherited from a superclass must include the data type of fields and the data types for the returned value and for the formal arguments of methods. The call to `exposeClass()` will typically have a named list for the `fields` and `methods` arguments. Named elements of the list will be a character vector specifying the corresponding C++ data types.

Let's look at an example, a class that extends the previous "PopBD" class:

```
#include "PopBD.h"

class PopCount: public PopBD {

  public:
    PopCount(void);
    PopCount(NumericVector initBirth, NumericVector initDeath);

    std::vector<long> table();

};
```

The new class adds one method, `table()`, that counts the current members of the population for each element in the lineage (very simple in C++).

We can share this class with R, omitting the shared `lineage` field in class `PopBD` in favor of its summary via the `table()` method. (As with real populations, the population size can grow exponentially, with `lineage` growing accordingly. The `table()` method keeps the data exchange with R to a constant size.) Here is the corresponding call to `exposeClass()`:

```
> exposeClass("PopCount",
+       constructors =
+         list("", c("NumericVector", "NumericVector")),
+       fields = list( size = "std::vector<long>"),
+       methods = list(evolve = c("void", "int"), "table"),
+       header = '#include "PopCount.h"', readOnly = "size")
Wrote C++ file "src/PopCountModule.cpp"
Wrote R file "R/PopCountClass.R"
```

The inherited `size` field is specified with its data type, and again made read-only. The inherited `evolve()` method is given with its return type (`void`) and the type(s) of its arguments, in this case just one (`int`). The `table` method is direct, not inherited, so we only need to supply its name.

For the same data as before:

```
> birth <- rep(c(.04,.06),10)
> death <- rep(.02, 20)
> p1 <- PopCount(birth, death)
> p1$evolve(100)
> ## make a table for the two birth probabilities
> tbl <- matrix(p1$table(), nrow = 2)
> rowSums(tbl)
[1]   72 351
```

If we continue to evolve this population, the size passes 1 million after 311 generations, so the efficiency of C++ may become relevant.

To experiment with extensions or modifications of the class, one can define a reference class that contains the proxy class, adding or overriding methods or fields. However, note that the proxy class is defined by a load action in order to use information from the C++ code. In the loading steps for a package outlined in Section 7.3, dynamically linked libraries are loaded into the process before the load actions are carried out.

A regular R reference class that subclasses a proxy class in the same package needs to be defined in a call to `setRefClass()` from a load action. That load action must take place after the load action for the proxy class itself, which comes from the R file generated by `exposeClass()`. If the subclass is defined in the same package as the proxy class, its load action needs to be set in a source file that comes after the generated file, either by using an explicit `Collate` directive or by relying on alphabetical order. The generated R file for the proxy class has the name of the class; for example, `"PopCountClass.R"` in our second example above.

The whole `exposeClass()` mechanism is not as straightforward as we would like, compared either to the `compileAttributes()` mechanism for proxy C++ functions or to the proxy classes in Chapters 14 and 15. In terms of the desirable pyramid mentioned in discussing XR—the work needed for interface designers, application designers and end users—the `exposeClass()` mechanism puts too much burden on the middle of the pyramid.

A better mechanism likely requires access to the information constructed by an actual C++ compiler to describe the class. C++ has no metadata objects for this purpose. R, Python and Julia do, and we used those for proxy class definitions. The existence of multiple C++ compilers and the continuing evolution of the language complicate the project, but a future contribution along these lines would be a boost to extending R.

# 16.5 Data Conversion

Data conversion in Rcpp differs from that in the XR interfaces in that it is essentially all programmed in the server language. Specifically, it uses two C++ functions defined by templates parametrized by typename:

```
template <typename T> T as( SEXP x) ;

template <typename T> SEXP wrap(const T& object) ;
```

If a particular C++ data type occurs as an input argument to a C++ function or method call or as the type of a field being assigned from R, there must be a valid templated version of as() available to Rcpp to convert an R object. Similarly, for the value of a function call or of an accessed field in a C++ object to be returned from a particular data type, there must be a templated version of wrap() to convert object to an R object.

The template programming requirement is less intimidating than it might seem. Such definitions exist for a wide range of data types, both within Rcpp itself and in the many packages that use it to interface to other software. In addition, any package using Rcpp can extend classes that the package owns by two additions to the class definition.

1. Providing a constructor for the class that takes a general R object as its argument (type "SEXP") will enable the generic version of as() to work.

2. Similarly, implementing and declaring a conversion operator to a general R object enables the generic wrap() to deal with this class; i.e., a method for operator SEXP().

To provide for a class that the package does *not* own requires some template programming to implement specialized definitions of as() and wrap() to support new data types. The extensions vignette [16] in the Rcpp package describes the procedures.

Rcpp covers a number of commonly occurring types. Here are some of the classes provided for:

- C++ classes corresponding to R vectors and matrices. The naming convention for these is "Vector" or "Matrix", with the data type pasted on the front, initial caps, for example "NumericVector" in the function above. Except that "List" was used instead of "ListVector" for R vectors of type "list".

- C++ classes corresponding to other basic R data types and classes. Many of the objects likely to be useful in interfaces are valid. The naming scheme is usually the data type (the value of `typeof()`) with initial capital, for example `"Symbol"` for class `"name"` and `"Language"` for all the language classes. An important exception is `"Function"` for R functions, which Rcpp deals with quite effectively.

  Some commonly used S3 classes are also included, such as `"Formula"`. Some of the data types have relatively few methods in C++, but still are useful for type checking and argument passing.

- The C++ vector classes and other classes in the Standard Template Library. The standard library defines a templated set of classes for vectors containing various types of data. These "vectors" are dynamically resizable arrays with methods for iterations and other common basic operations; for example, `std::vector<double>` is a vector of numeric values. Rcpp will map these objects to and from R objects that can be interpreted as numeric vectors.

Additional classes may be provided by packages among the many hundreds that depend on Rcpp on CRAN and elsewhere.

Objects that are pointers to particular data types can be declared by the `XPtr<T>` parametrized type. R has the `"externalptr"` type to hold general pointers or references, but objects of this type have no information about *what* they point to. The Rcpp templated type allows conversions to and from `"externalptr"` with computations in the C++ code specialized to the type `T` of the object referred to. Templated code can in fact use 3 parameters: type, storage scope and a finalizer to be called when the object pointed to is deleted. The object returned to R is in all cases of class and type `"externalptr"`. No information is available about the parametrized form.

For data types that are converted to or from ordinary R objects, particularly vectors, care needs to be taken when overwriting parts of the object. C++ and the `.Call()` interface generally will not copy references to R vectors. For the basic types `"NumericVector"` and the like, `copy()` should be used before any elements are modified to ensure that R objects are not corrupted. With other C++ types that can hold vector-like data, investigate carefully before overwriting data without copying.

# Bibliography

[1] J.J. Allaire, Dirk Eddelbuettel, and Romain François. *Rcpp Attributes*. Vignette in the Rcpp package.

[2] Richard A. Becker and John M. Chambers. GR-Z: A system of graphical subroutines for data analysis. In *Proc. 9th Interface Symp. Computer Science and Statistics*, 1976.

[3] Richard A. Becker and John M. Chambers. *S: An Interactive Environment for Data Analysis and Graphics*. Wadsworth, Belmont CA, 1984.

[4] Richard A. Becker and John M. Chambers. *Extending the S System*. Wadsworth, Belmont CA, 1985.

[5] Richard A. Becker, John M. Chambers, and Allan R. Wilks. *The New S Language*. Chapman & Hall, Boca Raton, FL, 1988.

[6] Carlos J. Gil Bellosta. *rPython: Package allowing R to call Python*. R package https://CRAN.R-project.org/package=rPython.

[7] Matthias Burger, Klaus Juenemann, and Thomas Koenig. *RUnit: R Unit Test Framework*, 2015. R package https://CRAN.R-project.org/package=RUnit.

[8] John M. Chambers. *Computational Methods for Data Analysis*. John Wiley and Sons, New York, 1977.

[9] John M. Chambers. Interface for a Quantitative Programming Environment. In *Comp. Sci. and Stat., Proc. 19th Symp. on the Interface*, pages 280–286, March 1987.

[10] John M. Chambers. *Programming with Data: A Guide to the S Language*. Springer, New York, 1998.

[11] John M. Chambers. *Software for Data Analysis: Programming with R.* Springer, New York, 2008.

[12] John M. Chambers and Trevor Hastie, editors. *Statistical Models in S.* Chapman & Hall, Boca Raton, FL, 1992.

[13] Winston Chang. *R6: Classes with reference semantics.* R package `https://CRAN.R-project.org/package=R6`.

[14] Wilfrid R. Dixon, editor. *BMDP Statistical Software.* University of California Press, 1983.

[15] Dirk Eddelbuettel. *High Performance and Parallel Computing.* `http://cran.r-project.org/web/views/HighPerformanceComputing.html`.

[16] Dirk Eddelbuettel and Romain François. *Extending Rcpp.* Vignette in the `Rcpp` package.

[17] Michael Fogus. *Functional JavaScript: Introducing Functional Programming with Underscore.js.* O'Reilly Media, 2014.

[18] HaskellR. *Programming R in Haskell.* URL `http://tweag.github.io/HaskellR/`.

[19] James Honaker, Gary King, and Matthew Blackwell. Amelia II: A program for missing data. *Journal of Statistical Software,* 45(7):1–47, 2011. URL `http://www.jstatsoft.org/v45/i07/`.

[20] Jeffrey Horner. *Rook: Rook - a web server interface for R.* R package `https://CRAN.R-project.org/package=Rook`.

[21] *SOAP II - Symbolic Optimal Assembly Program for the IBM 650 Data Processing System.* IBM, 1957. `http://bitsavers.trailing-edge.com/pdf/ibm/650/24-4000-0_SOAPII.pdf`.

[22] Ross Ihaka. *R : Past and Future History,* 1998. (draft for Interface Symp. Computer Science and Statistics): `https://cran.r-project.org/doc/html/interface98-paper/paper.html`.

[23] Ross Ihaka and Robert Gentleman. R: A language for data analysis and graphics. *Journal of Computational and Graphical Statistics,* 5:299–314, 1996.

[24] Kenneth E. Iverson. *A Programming Language.* Wiley, 1962.

[25] Jessica Lanford(ed.), Spencer Aiello, Eric Eckstrand, Anqi Fu, Mark Landry, and Patrick Aboyoun. *Machine Learning with R and H2O.* www.h2o.ai/resources.

[26] C. Lattner and V. Adve. Llvm: A compilation framework for lifelong program analysis and transformation. In *Proc. of the 2004 International Symposium on Code Generation and Optimization (CGO04)*, page 7588, San Jose, CA, USA, 2004. http://llvm.org/.

[27] Deborah Nolan and Duncan Temple Lang. *XML and Web Technologies for Data Sciences with R.* Springer, New York, 2014.

[28] Jeroen Ooms. *V8: Embedded JavaScript Engine.* R package https://CRAN.R-project.org/package=V8.

[29] Jeroen Ooms. The jsonlite Package: A Practical and Consistent Mapping Between JSON Data and R Objects. *http://arxiv.org/abs/1403.2805 [stat.CO]*, 2014.

[30] R Special Interest Group on Databases. *DBI: R Database Interface.* R package https://CRAN.R-project.org/package=DBI.

[31] rpy2. *R in Python.* URL http://rpy.sourceforge.net/.

[32] Andrew Shalit. *The Dylan Reference Manual.* Addison-Wesley Developers Press, Reading, MA, 1996.

[33] Duncan Temple Lang. Compiling code in R with LLVM. *Statistical Science*, 29(2):181–200, 2014.

[34] Luke Tierney. Name Space Management for R. *R News*, 3(1):2–6, June 2003.

[35] Luke Tierney. A Byte Code Compiler for R. Technical report, University of Iowa, October 2014. URL http://homepage.stat.uiowa.edu/~luke/R/compiler/compiler.pdf.

[36] Hadley Wickham. *Advanced R.* Chapman & Hall/CRC, 2014.

[37] Hadley Wickham. *R Packages.* O'Reilly, 2015.

[38] Hadley Wickham, Peter Danenberg, and Manuel Eugster. *roxygen2: In-Source Documentation for R,* . R package https://CRAN.R-project.org/package=roxygen2.

[39] Hadley Wickham, David A. James, and Seth Falcon. *RSQLite: SQLite Interface for R,* . R package https://CRAN.R-project.org/package=RSQLite.

# Index

CPI Group (UK) Ltd, Croydon, CR0 4YY
23/10/2024
`7692-0012